Counting Statistics for Dependent Random Events

Enrico Bernardi • Silvia Romagnoli

Counting Statistics for Dependent Random Events

With a Focus on Finance

Springer

Enrico Bernardi
Department of Statistical Sciences
"Paolo Fortunati"
University of Bologna
Bologna, Italy

Silvia Romagnoli
Department of Statistical Sciences
"Paolo Fortunati"
University of Bologna
Bologna, Italy

ISBN 978-3-030-64249-5 ISBN 978-3-030-64250-1 (eBook)
https://doi.org/10.1007/978-3-030-64250-1

Mathematics Subject Classification: 62P05, 62H20, 91G40, 60C05, 11P81, 05A17, 60B20

This Springer imprint is published by the registered company Springer Nature Switzerland AG.
The registered company address is: Gewerbestrasse 11, 6330 Cham, Switzerland

Preface

This book presents a conclusive synthesis of a number of years of original research carried out at the University of Bologna into some probabilistic problems involving the interaction of the dependence structure and combinatorics-based algorithms.

The questions tackled here arise directly from applications where one is frequently confronted with issues linked to a counting variable monitoring a random event. For instance, these can happen in credit applications, where one may imagine to control the defaults attached to a portfolio of loans, or they can occur in insurance problems, where one needs to monitor the claims' arrivals in order to detect the suitable level of risk the company can bear.

This rather general research field belongs to the area of counting statistics or, equivalently, to the domain of statistics of counting random events, historically encompassing primarily the study of point processes. As it is well-known a point, process in time is a stochastic process where a particular type of event is classified only according to a single point on a time line, for example, the variation in time of the births of babies in a particular hospital, the passage of cars past a counting gate on a toll bridge, or the arrival of particles from outer space at a detector. Such point processes may be further categorized as Poisson processes if the occurrence of any particular point is completely independent of the occurrence of any other point. A Poisson process considered on the positive half-line can be defined as a counting process representing the total number of occurrences or events that have happened up to a given time. On the other, hand if the criterion of independence is not satisfied, the point process is more generally known as a renewal process.

Our new plan in this field sets out from a very general framework where neither independence nor identical distribution assumptions are taken into account. We analyze the phenomena at first from a static point of view, where, as per the classical outline, the point process falls into a single random variable which is a counting variable as well. Such a random variable expresses the probability of a given number of events occurring in a fixed interval of time, if these events happen with a known constant rate and independently of the time since the last event. This observation may seem trivial but allows us to simplify the discussion considering at first the problem at a fixed point in time and including his evolution in time only later.

Let us consider the static problem of counting; the well-known results on counting processes usually rely on the assumption of independence of occurrences, and this is not representative, in general, of the most common real phenomena. Hence, the idea is to generalize the problem assuming that we are interested in counting the occurrences of a random event which may involve a set of dependent objects/individuals, and this last one may be very big indeed. Those dependences must impact the calculus of the joint probabilities concerning every state of the world and also the arrival times of the random events themselves. These two kinds of impacts characterize our new technique explained in all detail in this book. The main ingredients of our algorithm are: the combinatorics, the clustering, and the dependence structure. As one can imagine, the combinatoric distributions of the random event on the set of objects/individuals enable us to detail the set of all possible scenarios. Nevertheless, a strictly linked issue concerns the impact of the dependence structure on the combinatoric algorithm and on the selection of the most probable scenarios. As it will be made clear later, this selection leads to the assignment of a weight, whose meaning looks like a degree of confidence on the scenario at hand. Moreover, empirical feedback gives us instructions about the kind of dependence that seems to be more likely the real one, highlighting the relevance to consider hierarchical dependence structure in order to introduce the phenomena of contagions. Contagion is strictly related to the dependence structure and can be considered at two levels: the highest one refers to the systemic risk, while the lowest one is associated to the influence among similar objects/individuals that could be grouped in homogeneous classes. The degree of homogeneity is evaluated through a clustering approach capable to shed light on the concept of similarity and to decrease the level of complexity of the problem. Indeed, also in case of data sets that are so large or complex that they need advanced data analytics methods, we are able to make the problem tractable through a sophisticated clustering approach. Taking into account all of the available information, we identify groups of variables sharing common characteristics and similarities and we focus our attention to the dependence structure among groups rather than the single object/individual. A last caveat is in order about the way we model the dependence structure in our novel algorithmic approach. This dependence is relevant in two main phases of the algorithm, i.e,. in the combinatoric distribution of the random event to detail the scenarios, and in the aggregation phase related to every state of the world. For this purpose, we define a new class of copulas, called *clusterized copulas*, since they are based and modeled on these important features of the algorithm. In fact those copulas describe the dependence structure among groups based on a preliminary clustering of objects/individuals, accounting for possible hierarchy of groups, and provide the joint probability of every scenario detailed by *combinatorics tools*. From this point of view, the definition of a clusterized-based dependence structure on which a combinatoric algorithm could be integrated can be seen as a generalization and an alternative to an ideal mix of an intensity-based model and of a copula convolution; this is an answer to a general representation of an ever more complex and intricate world.

The dynamic version of our algorithm implies the switching from a static representation of the dependence structure to a dynamical one, where the dependence of levels and increments should be specified, in line with the concept of dynamic copulas. Therefore, the representation of the process in discrete time is, in this context, strictly related to the evolution of the discriminant properties of objects/individuals for clustering. The knowledge of the discriminant properties' evolution and then of the new clustering organization at every period in the future implicitly defines the dynamic of our aggregation algorithm. Nevertheless, the main features of the dynamic algorithm are the same as previously described in a static context.

Hence, a graphic representation of the main step of the algorithm may be useful to the reader in order to understand the sequential logic of every chapter and to appreciate their connection along with the final goal of the writing.

The outline of the book is as follows. In the first part, we will introduce the main instruments of our algorithm, that is, the clustering, the combinatorics, and the dependence structure. In this first part, the reader can understand the new perspectives presented later in the second part: here, the fundamental concepts are displayed in a way that is more suitable to our further purposes. Therefore, we will start considering a set of objects/individuals characterized by several features. Those features are selected coherently with the problem at hand, and based on similarity criteria, the complexity of the problem is then reduced by a clustering; from now on, the focus will be on the groups rather than on the single object/individual. Every group is a set of homogeneous objects/individuals sharing their selected features and the precision will be improved greater in the number of groups. The best choice of the number of clusters is an important issue that will be discussed later on.

Based on the new set organized in groups as depicted in Fig. 1, and with the knowledge of their dependence structure, we move on to the next step which involves the recovering of all possible scenarios related to the distribution of a random event into the groups. Finally, this cdf can be retrieved using the aggregation's instructions provided for in Part II. Indeed, in the second part, the ingredients of Part I will be arranged in such a way that the resulting new algorithm will be ready for a detailed analysis. Several applications based on real data will be proposed eventually, and the performance comparison of the new algorithm to alternative approaches already proposed in the literature will be considered and further analyzed.

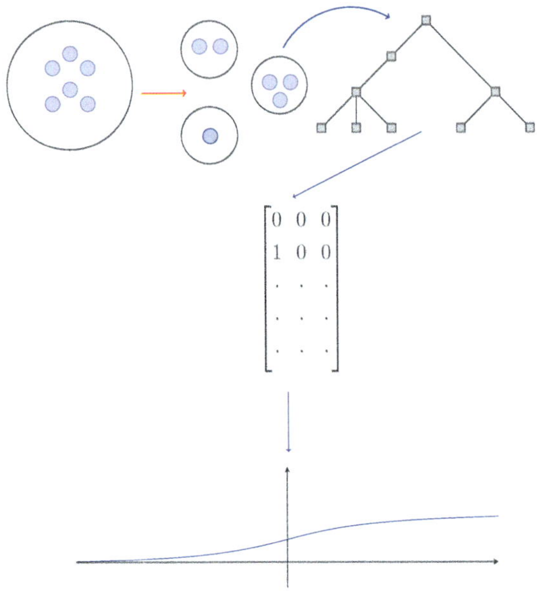

Fig. 1 Graphical Structure of the main step of the Aggregation Algorithm

It is our belief to contribute with this book to the existing literature either in the field of counting statistics and in the more general applied mathematics area, with a special focus to the financial applications. Therefore, we figure out to reach the theoretical interest of master 's and doctoral students for the innovative methodologies put forward in this book but also to stimulate the empirical employment of new approaches from professionals and practitioners.

Bologna, Italy

Enrico Bernardi
Silvia Romagnoli

List of Common Symbols, Notations, and Acronyms

$\square$ end of proof
$\mathbb{N}$ the set of natural numbers
$I = [0, 1]$ the unit interval of the real line
$\mathbb{R} = (-\infty, +\infty)$ the real line
$\mathbb{R}^+ \setminus \{0\} = (0, +\infty)$ the positive real line
$[a, b] \times [c, d]$ cartesian product of the intervals $[a, b]$, $[c, d]$
$\mathbb{R}^n = \underbrace{(-\infty, +\infty) \times (-\infty, +\infty) \times \ldots \times (-\infty, +\infty)}_{n \text{ times}}$ the $n-$ dimensional Euclidean vector space
$I^n = \underbrace{[0, 1] \times [0, 1] \times \ldots \times [0, 1]}_{n \text{ times}}$ unit cube in R^n, $n \geq 2$
$\mathbf{x} = [x_1 x_2 \ldots x_n]^T$ n-dimensional (column) vector
$\mathbf{x}' =$ the transpose of the vector $\mathbf{x}$
C copula function
ϕ generator of an archimedean copula
$\phi^{[-1]}$ pseudo-inverse of ϕ
$F(x, y)$ bivariate distribution function (cumulative probability function) of the random vector $[X, Y]$, computed at (x, y)
F_i univariate distribution function (cumulative probability function) of the $i - th$ random variable
F_i^{-1} generalized inverse of F_i
Ran F is its range
C^∞ the space of functions $f : R \to R$ with derivatives of all orders
L^2 is the space of random variables with finite first two moments
C^+ upper Fréchet-Hoeffding bound
C^- lower Fréchet-Hoeffding bound (copula for $n = 2$)
$C^\perp$ product copula
$\bar{C}$ survival copula
$H \overset{C}{*} F$ C-convolution operator
$VaR_\alpha(X)$ value-at-risk of an exposure X at confidence level α
$\mathbf{1}_{\{E\}}$ indicator function of the event E
$sgn(x) = \begin{cases} -1 & x < 0 \\ 0 & x = 0 \\ 1 & x > 0 \end{cases}$ signum function
wrt = with respect to
rv = random variable
iff = if and only if
cdf = cumulative distribution function
pdf = probability distribution function
c.d.= combinatorial distribution
c.c.d. = compatible (with respect to group's cardinality) combinatorial distribution

o.c.c.d. = ordered compatible combinatorial distribution
a.e. = except than in a set of Lebesgue measure zero
CHC = Clusterized Homogeneous Copula
CHY = Clusterized Hierarchical Copula
DHC = Distorted Hierarchical Copula
HLC = Hierarchical Limiting Copula
HLY = Hierarchical Hybrid Copula
PL = Pure Limiting Model
PE-HC = Partially Exchangeable Hierarchical Copula
NE-HC = Not Exchangeable Hierarchical Copula

Contents

Part I
The Main Ingredients

Abstract A nonconventional review of the main ingredients of a new aggregation algorithm, i.e., the clustering, the combinatorics, and the dependence structure, will allow the reader to understand the new perspectives discussed in Part II.

Chapter 1
Clustering

Abstract The goal of this chapter is to survey and present the main concepts and techniques from the vast collection of clustering models, from the hierarchical to the partitioning ones. In the partitioning class, we extensively discuss the clustering approaches that can be seen as particular neural networks, with attention to the key concepts involved into the social sciences' applications. The clustering is one of the main ingredients of the new copula-based algorithm of aggregation which will be discussed in Part II.

Researchers in many disciplines frequently face the difficulty to work with enormous data sets whose characteristics must be analyzed and synthesized so as to allow a deeper analysis. Data analysis aims to explore interrelationships among data points to make an assessment of their structure, but if the dimension of the problem is too high, we need techniques able to reduce this complexity.

A key element of data analysis is represented by the cluster analysis which is quite valuable since it can drastically reduce the dimensionality of the data. The goal of clustering is to identify a structure in an unlabelled data set by objectively organizing data into homogeneous groups where the within-group-object similarity is maximized and the between-group-object dissimilarity is maximized. Hence cluster analysis allows to organize a collection of patterns (observations, data items, or feature vectors) into clusters based on similarity.

The variety of techniques for representing data, measuring similarity between elements, and then grouping data elements has produced a rich collection of clustering methods mainly depending on the characteristics of data. If their feature values do not change with time or change negligibly, we have static data. The bulk of clustering analysis has been performed on static data. On the other hand, if data are time series, the existing algorithms for clustering static data are generalized in order to be directly used also for time series clustering.

As explained in the Preface, clustering is one of the main ingredients of the new algorithm of aggregation that will be introduced in the next sections. Therefore we will proceed now with a short review of the fundamentals in clustering.

E. Bernardi, S. Romagnoli, *Counting Statistics for Dependent Random Events*,
https://doi.org/10.1007/978-3-030-64250-1_1

However, a comprehensive overview of clustering models, which have a rich history in other disciplines, such as biology, psychiatry, psychology, archaeology, geology, geography, and marketing, is beyond the scope of this presentation, and for more details we refer, e.g., to Gallant [7] and Jain et al. [10].

1.1 Preliminary on Clustering

Typical clustering model aggregates data into several classes based on a given set of input features that is a set $\mathcal{X}$ of d-dimensional vectors $\mathbf{x}_i = (x_i^1, \ldots, x_i^d)$ such that

$$\mathcal{X} = \{\mathbf{x}_1, \ldots, \mathbf{x}_n\}.$$

This way input features correspond to a $n \times d$ pattern matrix.

Clustering process involves the following steps:

(i) Pattern representation. This includes the feature *selection* and the feature *extraction* processes. Feature *selection* aims to identify the most effective subset of original features to use in clustering, while feature *extraction* refers to the use of transformations of the input features to produce new more appropriate features for the clustering at hand.

(ii) Definition of a pattern proximity measure. This is usually a distance function defined on pairs of patterns, but other choices could be **made** as we'll see in the following section.

(iii) Clustering. This step can be performed in a number of ways. The output clustering is *hard* if a class label l_i with $l_i \in \{1, \ldots, k\}$ and where k is the number of clusters is assigned to each pattern $\mathbf{x}_i$ in $\mathcal{X}$ identifying its class. On the other hand, a *fuzzy* clustering assigns to each pattern $\mathbf{x}_i$ in $\mathcal{X}$ a fractional degree of membership f_{ij} in each output cluster j, for $j = 1, \ldots, k$.

Moreover we can distinguish between *partitioning* approaches, identifying the partition of data that optimizes (usually locally) a clustering criterion and *hierarchical* ones producing a nested series of partitions based on merging and splitting clusters depending on similarity.

(iv) Cluster validity analysis. It determines whether the output is meaningful. There are three types of validations: the *external* test compares the recovered structure to an priori structure; the *internal* test determines if the structure is intrinsically appropriate for the data; and the *relative* test compares two structures and measures their relative merit.

Since the choice of the similarity measure plays a central role in clustering model setup, we analyze in details the main solutions proposed in literature, in the following section.

1.2 The Similarity Measure for Static Data

A measure of similarity between two patterns drawn from the same feature space is essential to most clustering procedures. It is most common to calculate the dissimilarity between two patterns using a distance measure defined on the feature space.

Metric or quasi-metric selection is necessary to establish a rule for assigning patterns to the domain of a particular cluster. As expected, the most popular metric for continuous features is the *Euclidean distance*

$$d_2(\mathbf{x}_i, \mathbf{x}_j) = \|\mathbf{x}_i - \mathbf{x}_j\|_2.$$

Euclidean distance is one of the most common *Minkowski distance* metrics defined as

$$d_p(\mathbf{x}_i, \mathbf{x}_j) = \|\mathbf{x}_i - \mathbf{x}_j\|_p.$$

This choice works well when data sets are characterized by compact and isolated clusters. In fact the main drawback of the direct use of Minkowsky measures is the tendency of the largest-scaled feature to dominate the others. To solve this problem, we need to normalize the continuous features to a common range of variance.

A possible distortion of data comes also from the linear correlation among them; this kind of distortion can be alleviated by choosing as metric the *squared Mahalanobis distance*, defined as

$$d_M(\mathbf{x}_i, \mathbf{x}_j) = (\mathbf{x}_i - \mathbf{x}_j)\Sigma^{-1}(\mathbf{x}_i - \mathbf{x}_j)^\top,$$

where Σ is the sample covariance matrix of the patterns. We observe that Mahalanobis distance assigns different weights to different features based on their variances and pairwise linear correlations. Here we implicitly assume that the density function is a multivariate Gaussian.

Computation of distances between patterns with some noncontinuous feature is hard since the feature cannot be compared. Wilson and Martinez [27] propose a proximity measure for heterogeneous patterns using a modified Minkowsky metric for continuous features and a distance based on counts for nominal attributes for noncontinuous ones.

There are other distances proposed in literature which take into account the effect of surrounding or neighboring points called *context*. The similarity between two points, given the context $\mathcal{E}$, is defined as

$$s(\mathbf{x}_i, \mathbf{x}_j) = f(\mathbf{x}_i, \mathbf{x}_j, \mathcal{E}).$$

One metric, defined using the *context*, is the *mutual neighbor distance* (MND in the following) that has been proposed in Gowda and Krishna [8] and is defined as

$$d_{MND}(\mathbf{x}_i, \mathbf{x}_j) = NN(\mathbf{x}_i, \mathbf{x}_j) + NN(\mathbf{x}_j, \mathbf{x}_i),$$

where $NN(\mathbf{x}_i, \mathbf{x}_j)$ is the neighbor number of $\mathbf{x}_j$ with respect to $\mathbf{x}_i$. We point out that even though MND is not a metric, it has been successfully applied in cluster analysis, supporting the viewpoint that the dissimilarity doesn't need to be a metric.

In the same line, Su and Chou [25] proposed a new nonmetric measure, based on the idea of symmetry which can be used to group a given data set into a set of clusters of different geometrical structures. In Wang et al. [26], a new data-dependent dissimilarity measure was designed, called *density-sensitive distance* metric, which has the property of elongating the distance between points in different high-density regions and then can reflect the characters of data clustering.

Other contributions have focused more on the performance of a metric in a randomized data set, arguing that in the case of absence of noise of any kind, virtually all metrics are equivalent. In Francois et al. [6], it is suggested to choose the metric according to the shape of the noise that is assumed on the data. While it is known that the Euclidean metric is optimal in presence of white Gaussian noise, other types of noise may require other metrics.

1.3 The Similarity Measure for Time Series

In the special context of time series data, the concept of dissimilarity must take into account the dynamic character of the series. Dissimilarities usually considered in static clustering ignore the interdependence relationship between values which cannot be disregarded in time series clustering. For this reason different approaches to define dissimilarities between time series have been proposed in literature as detailed in Liao [16] and Montero and Vilar [19]. A short overview is presented below.

The dissimilarity measures for time series can be classified mainly into two categories: model-free measure and model-based measures. We consider in the following the T-dimensional vector of realizations of two real-valued processes X and Y, which are noted as $\mathbf{X}_T$ and $\mathbf{Y}_T$.

1.3.1 Model-Free Approaches

The aim is to measure the proximity between $\mathbf{X}_T$ and $\mathbf{Y}_T$ with a metric based on the closeness of their values at specific points of time.

The *Minkowski distance* and the special case of *Euclidean distance* can be used also in this context, i.e.,

$$d_p(\mathbf{X}_T, \mathbf{Y}_T) = \|\mathbf{X}_T - \mathbf{Y}_T\|_p.$$

Here the proximity notion relies on the closeness of values observed at corresponding points of time so that the observations are assumed to be independent. Unlike the *Minkowski distance*, the *Fréchet distance*, i.e.,

$$d_F(\mathbf{X}_T, \mathbf{Y}_T) = \min_{r \in M} \left(\max_{i=1,\dots,m} |X_{a_i} - Y_{b_i}| \right),$$

where M denotes the set of all possible sequences of m pairs preserving the observation order and $a_i, b_i \in \{1, \dots, T\}$, where $a_1 = b_1 = 1$ and $a_m = b_m = T$ does not treat the series as two points sets but takes into account the ordering of the observations. This distance allows to recognize similar shapes but ignores the temporal structure of the values. A dissimilarity index model including both behavior and values proximity estimation is based on the first order temporal correlation coefficient which is defined by

$$CORT(\mathbf{X}_T, \mathbf{Y}_T) = \frac{\sum_{t=1}^{T-1} (X_{t+1} - X_t)(Y_{t+1} - Y_t)}{\sqrt{\sum_{t=1}^{T-1} (X_{t+1} - X_t)^2} \sqrt{\sum_{t=1}^{T-1} (Y_{t+1} - Y_t)^2}}.$$

It belongs in the interval $[-1, 1]$; if it is 1, the series show the same dynamic behavior meaning that at any instant their evolution is similar in direction and rate, while if it is -1 the series have a similar growth in rate but opposite direction. The independence is represented by a zero value. The corresponding dissimilarity index is given by

$$d_{CORT}(\mathbf{X}_T, \mathbf{Y}_T) = \phi_k(CORT(\mathbf{X}_T, \mathbf{Y}_T)) \times d(\mathbf{X}_T, \mathbf{Y}_T),$$

where $\phi_k(u) = \frac{2}{1+\exp(ku)}$, $k \leq 0$ is an adaptive tuning function which modulates a raw data distance $d(\mathbf{X}_T, \mathbf{Y}_T)$ according to the temporal correlation. In the same spirit, we may think to replace raw data by a reduced number of features characterizing the time series in order to consider the temporal structure of the series. This way we define the correlation-based distances among which we have the metric depending on the Pearson's correlation, i.e.,

$$COR(\mathbf{X}_T, \mathbf{Y}_T) = \frac{\sum_{t=1}^{T} (X_t - \overline{\mathbf{X}}_T)(Y_t - \overline{\mathbf{Y}}_T)}{\sqrt{\sum_{t=1}^{T} (X_t - \overline{\mathbf{X}}_T)^2} \sqrt{\sum_{t=1}^{T} (Y_t - \overline{\mathbf{Y}}_T)^2}},$$

where $\overline{\mathbf{X}}_T, \overline{\mathbf{Y}}_T$ stand for the mean values of the series. The corresponding cross-correlation-based distances were suggested:

$$d_{COR.1}(\mathbf{X}_T, \mathbf{Y}_T) = \sqrt{2(1 - COR(\mathbf{X}_T, \mathbf{Y}_T))}$$

$$d_{COR.2}(\mathbf{X}_T, \mathbf{Y}_T) = \sqrt{\left(\frac{1 - COR(\mathbf{X}_T, \mathbf{Y}_T)}{1 + COR(\mathbf{X}_T, \mathbf{Y}_T)}\right)^{\beta}}, \beta \leq 0.$$

Sometimes it may be convenient to use a clustering based on the estimated autocorrelation vectors computing the following distance:

$$d_{ACF}(\mathbf{X}_T, \mathbf{Y}_T) = \sqrt{(\hat{\rho}_{X_T} - \hat{\rho}_{Y_T})'\Omega(\hat{\rho}_{X_T} - \hat{\rho}_{Y_T})},$$

where Ω is a matrix of weights. If we assume $\Omega = I$, we recover the Euclidean distance between the estimated autocorrelation functions. Analogous distances can be constructed by considering the partial autocorrelation functions instead of the autocorrelation functions.

The metrics discussed above work in the time domain, but sometimes the frequency domain offers an interesting alternative to measure dissimilarity of time series. In this case we need to compute dissimilarities between the spectral representation of the series as suggested among other in the computation of the periodogram-based distances and the spectral disparity measures.

1.3.2 Model-Based Approaches

In this class we assume that the underlying models are generated from specific parametric structures. A common structure is the invertible ARIMA which needs to be estimated along with the parameters in order to compute a dissimilarity measure between the fitted models. The most relevant measures based on the assumption of underlying ARIMA models are the *Piccolo* and the *Maharaj* distances. If $\hat{\Pi}_{X_T} = (\hat{\pi}_{1,X_T}, \ldots, \hat{\pi}_{k_1,X_T})'$ and $\hat{\Pi}_{Y_T} = (\hat{\pi}_{1,Y_T}, \ldots, \hat{\pi}_{k_2,Y_T})'$ denotes the vectors of $AR(k_1)$ and $AR(k_2)$ parameter estimations for $\mathbf{X}_T$ and $\mathbf{Y}_T$, the *Piccolo* distance is given by the following:

$$d_{PIC} = \sqrt{\sum_{j=1}^{k}(\hat{\pi}_{j,X_T}* - \hat{\pi}_{j,Y_T}*)^2},$$

where $k = \max(k_1, k_2)$, $\hat{\pi}_{j,X_T}* = \hat{\pi}_{j,X_T}$ if $j \leq k_1$ and $\hat{\pi}_{j,X_T}* = 0$ otherwise and $\hat{\pi}_{j,Y_T}* = \hat{\pi}_{j,Y_T}$ if $j \leq k_2$ and $\hat{\pi}_{j,Y_T}* = 0$ otherwise.

On the other hand, the *Maharaj* distance is defined as follows:

$$d_{MAH} = \sqrt{T}(\hat{\Pi}_{X_T} * - \hat{\Pi}_{Y_T} *)' \hat{V}^{-1} (\hat{\Pi}_{X_T} * - \hat{\Pi}_{Y_T} *),$$

where $\hat{\Pi}_{X_T}*$, $\hat{\Pi}_{Y_T}*$ are the $AR(k)$ parameter estimations of $\mathbf{X}_T$ and $\mathbf{Y}_T$, respectively, $k = \max(k_1, k_2)$, and $\hat{V}$ is an estimation of $V = \sigma^2_{X_T} R^{-1}_{X_T}(k) + \sigma^2_{Y_T} R^{-1}_{Y_T}(k)$, where $\sigma^2_{X_T}$ and $\sigma^2_{Y_T}$ are the variances of the white noise processes associated with $\mathbf{X}_T$ and $\mathbf{Y}_T$ and R_{X_T} and R_{Y_T} denote the sample covariance matrix of the series. It is proved that d_{MAH} is asymptotically χ^2 under the hypothesis of equality of generating processes if it is based on the assumption of independent time series, while it is asymptotically χ^2 with k degrees of freedom in case of correlated processes.

In order to improve the flexibility of the model and accommodate the modelling of data with either continuous or noncontinuous domains, a number of mixture models and copula-based mixture models have been proposed. The use of copulas in model-based clustering offers two direct advantages over current methods: (i) the appropriate choice of copulas provides the ability to obtain a range of exotic shapes for the clusters, and (ii) the explicit choice of marginal distributions for the clusters allows the modelling of multivariate data of various modes in natural way. For details see among other Kosmidis and Karlis [14].

1.4 Hierarchical Algorithm

A distinction between *hierarchical* and *partitioning* approaches is really important so as to understand the different nature of these models. *Hierarchical* methods produce a nested series of partitions, while *partitioning* methods produce only one partition. Essentially the hierarchical approach is agglomerative[1] because it begins with each pattern in a distinct cluster and then merges clusters together until a stopping criterion is satisfied.

Hierarchical clustering groups data by creating a cluster tree or *dendrogram* that is a multilevel hierarchy where clusters at one level are joined as clusters at the next level. To implement a hierarchical clustering, the first step consists in finding the similarity or dissimilarity between every pair of objects in the data set by computing a dissimilarity matrix Y, whose elements are the distances between each pair of patterns. There are many ways to calculate this distance as we have seen in the previous section, for example, in static clustering through the *Euclidean distance* between the objects. Once the proximity between objects in the data set has been computed, the distance is used to determine the proximity of every object to each other such to construct the hierarchical cluster tree: the procedure links pairs of

[1] Divisive methods do the opposite of the agglomerative ones.

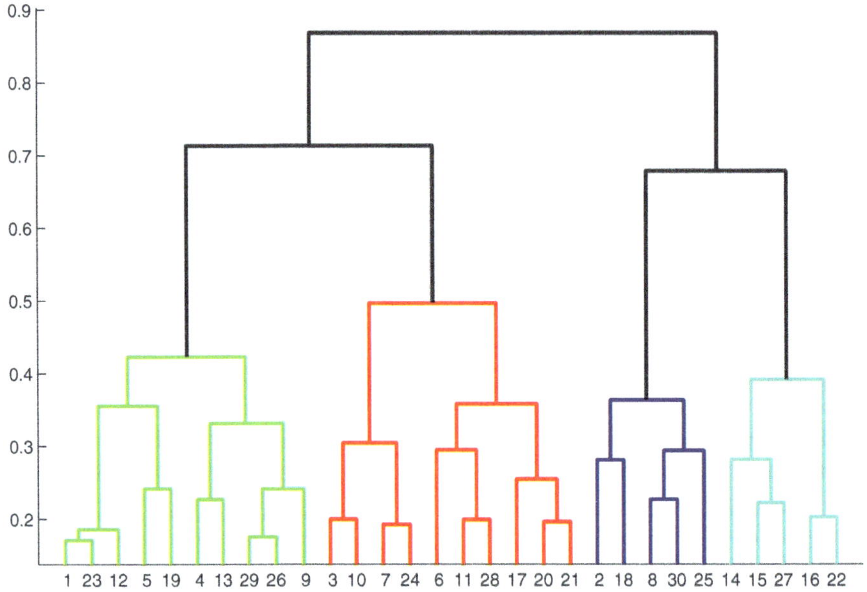

Fig. 1.1 Dendrogram generated by a data set with dimension 100 and t equal to 0.7 of the maximum distance

objects that are close together into binary clusters, and then the newly formed clusters are grouped into larger clusters until a hierarchical tree is formed. The cluster tree is easily understood when it is graphically represented as a *dendrogram* like that in Fig. 1.1. Here the horizontal axis reports the indices of the objects in the original data set, and the links between objects are represented as upside-down U-shaped lines whose height indicates the distance. The cluster tree is determined as a function of a distance measure which allows to link objects at some level; the height of the link (also called *linkage*) represents the distance between two clusters that contain those objects and defines the *cophenetic distance* between these objects. We observe that in Fig. 1.1 some leaf nodes correspond to multiple objects and different colors identify groups of nodes whose linkage is less than a threshold t fixed here as 0.7 of the maximum distance.

To verify that the distances, i.e., the heights, in the tree reflect the original distances, it is possible to compare the *cophenetic distances* with the original distance between data: the stronger the correlation between the linking and the distance between data, the better the performance of the clustering method. The *cophenetic correlation* between the cophenetic distances Z, representing the hierarchical clustering tree and the original distances Y of data, is defined as

$$cc = \frac{\sum_{i<j}(Y_{ij} - y)(Z_{ij} - z)}{\sqrt{\sum_{i<j}(Y_{ij} - y)^2 \sum_{i<j}(Z_{ij} - z)^2}},$$

where y and z are the original mean and the cophenetic distance. The *cophenetic correlation* allows to compare the results of clustering using different metric and linkage methods and to choose those best suited to represent the original distances.

The last step of the hierarchical method consists of creating a partition of data by taking into account the clustering tree. We can create these clusters by detecting natural grouping in the hierarchical tree or by cutting off the hierarchical tree at an arbitrary point. The natural cluster divisions in a data set is reached by comparing the height of each link in the cluster tree with the heights of neighboring links below it in the tree: a link with approximately the same height as the links below is said to be *consistent*, whereas a link whose height differs noticeably from the heights of the links below it is said to be *inconsistent*. The inconsistent links indicate the border of a natural division in a data set. The procedure uses a quantitative measure of inconsistency, i.e., an *inconsistency coefficient*, to determine where to partition the data set into clusters. To compute it, each link in the cluster hierarchy is compared with the adjacent links that are less than a fixed number of levels below it in the cluster tree; this integer is the *depth* of the comparison. The *inconsistency coefficient* of link k is defined as

$$ic(k) = \frac{Z(k) - MK(k)}{SDW(k)},$$

where $Z(k)$ is the *cophenetic distance* of link k, while $MW(k)$ and $SDW(k)$ are the mean and the standard deviation of the heights of all links included in the calculation once *depth* is defined.

Therefore the *inconsistency coefficient* characterizes each link in a cluster tree by comparing its height with the average height of other links at *depth* level below in the hierarchy. Obviously the leaf nodes, i.e., the objects at the bottom of the cluster tree, have no further objects below them and then clearly their inconsistency coefficients are zero. This way the inconsistency coefficients allow to naturally divide the data into distinct, well-separated clusters identified in the corresponding dendrogram by densely packed objects in certain areas and not in the others recalling that the higher the value of this coefficient, the less similar the objects connected by the link. Instead of a natural partition of data, it is possible to specify the desired number of clusters, which corresponds to applying a cut off value to the inconsistency coefficient.

Hierarchical clustering is also used to cluster time series not necessarily with equal length. Nevertheless a pure hierarchical method suffers from its inability to perform adjustments once a merge or split decision has been executed. To improve the performance of the clustering, usually it is integrated with other clustering techniques.

1.5 Partitioning Algorithm

A partitioning algorithm for clustering allows to obtain a single partition of the data instead of a clustering structure such as the dendrogram produced by the hierarchical method. Partitioning approaches work better than the hierarchical ones in applications involving large data sets, but they entail a problem concerning the choice of the number of desired output clusters.

Usually partitioning methods produce clusters by optimizing a criterion function defined either locally on a subset of the patterns or globally over all the patterns.

The squared error criterion is frequently used and tends to work well with isolated and compact clusters. The squared error for a clustering $\mathcal{L}$ of a pattern set $\mathcal{X}$ composed by k clusters is

$$e^2(X, \mathcal{L}) = \sum_{j=1}^{k} \sum_{i=1}^{n_j} \|\mathbf{x}_j^i - \mathbf{c}_j\|^2,$$

where $\mathbf{x}_j^i$ is the ith pattern in the jth cluster whose centroid, i.e., the mean element of group j, is $\mathbf{c}_j$. Basically the main steps in the implementation of the squared error clustering method are the following:

- Select an initial partition of the patterns with a fixed number of clusters and corresponding centroids.
- Assign each pattern to its closest centroid and repeat until the cluster membership is stable in term of no or minimal reassignment of patterns to new cluster or minimal decrease in square error.

k-means is the simplest and most commonly used algorithm employing a square error criterion, given the number of cluster to generate. In Bernardi and Romagnoli [3], the best number of groups given a subjective tolerance δ is defined as the minimum number of groups such that the maximum distance, computed on the chosen metric, of each data to the relative centroid, is not greater than the tolerance. In this way if the distance of a data is greater than the tolerance, one more group is made. Formally for one-dimensional patterns, they define the *δ-optimal number of groups*, n_δ, as the argument of the following optimization problem with a constraint:

$$n_\delta = \arg \min_{x_j^i - c_j \leq \delta, \forall i, j} CER,$$

where x_j^i, $\forall i, j$ represent the data, while c_j, $\forall j$ the set of centroids and the clustering error CER to minimize is

$$CER = \sum_{j=1}^{n_\delta} \sum_{i=1}^{n_j} (x_j^i - c_j)^2,$$

where n_j is the number of data in the jth group. We observe that CER is very similar to the clustering error defined as sum of *confusions* in Wang et al. [26].

Nevertheless the choice of the number of groups is a critical problem which involves the consideration of the trade-off between the complexity of the problem in terms of computing time and the precision of the method.

1.5.1 k-Means Clustering

One of the most popular partitioning clustering algorithms is k-means that has been proposed by MacQueen [17]. Several variations of this algorithm have been advanced; in the following we will refer to the convergent k-mean clustering version as described by Anderberg [1].

k-means clustering is a partitioning method, i.e., the function k-means partitions the data into k mutually exclusive clusters and returns a vector of indices pointing to the clusters which have been assigned to each observation.

It starts with a random initial partition and keeps reassigning the patterns to clusters based on the similarity between the patterns and the centroids until a convergence criterion is met. k-means treats each observation in a data set as an object having a location in space. It finds a partition in which objects within each cluster are as close as possible to each other and as far as possible from objects in other clusters. Each cluster in the partition is defined by its member objects and by its *centroid*, i.e., the point to which the sum of distances from all objects in that cluster is minimal. Usually we distinguish between k-means algorithm, where each cluster is represented by the mean value of the objects in the cluster, and k-medoids algorithm, where each cluster is represented by the most centrally located object in a cluster. Two counterparts for fuzzy partitions are the fuzzy c-means and the fuzzy c-medoids algorithms.

We recall that the *centroid* of the jth group, whose dimension is n_j, is the vector:

$$\mathbf{c}_j = \frac{1}{n_j}\sum_{i=1}^{n_j}\mathbf{x}_j^i.$$

k-means algorithm operates by moving an element to the cluster with the centroid closest to that one and then updating centroids for the revised clusters. The procedure uses an iterative algorithm that minimizes the sum of distances from each object to its cluster centroid, over all clusters. This algorithm moves objects between clusters until the sum cannot be decreased any further. The result is a set of clusters that are as compact and well-separated as possible. The algorithm is repeated until convergence is achieved that is until a pass through the data causes no new assignments.

Dunn [5] first extended the k-means algorithm to fuzzy partitioning, using the following objective function:

$$\min \sum_{i=1}^{k} \sum_{j=1}^{n} (\mu^i_j)^2 \|x_j - c_i\|^2,$$

where μ^i_j is an entry of the matrix of a fuzzy-partition. The constraints of the optimal problem are (1) $\mu^i_j \in [0, 1], \forall i, j$, (2) $\sum_{i=1}^{k} \mu^i_j = 1, \forall j$, and (3) $0 < \sum_{j=1}^{n} \mu^i_j < n, \forall i$. In other words, each x_j could belong to more than one cluster with each belongingness taking a value in [0, 1]. To run the procedure, we need to specify the number of clusters k.

A major problem is that k-means method is sensitive to the selection of the initial partition and may converge to a local minimum of the criterion function for an improper initial partition. Several variations were proposed in order to select a good initial partition so as to reach the global minimum value. Among these we recall the variation to permit splitting and merging of the resulting clusters so as to obtain the optimal partition starting from any arbitrary initial partition, provided proper threshold values for the max/min distance between the centroids are specified.

While the basic k-means clustering algorithm requires that a distance metric and the number of clusters, k, be chosen in advance, the algorithm is fast enough so that it is easy to experiment with different metrics and values of k.

Another approach is to let the algorithm determine the number of clusters in accordance with user-specified rule and parameters, given a distance metric (see Wishart 28). Several papers address the problem of clusters' number selection by using a k-means approach that exploits local changes of internal validity indices to split or merge clusters. In Pelleg and Moore [23] this issue is addressed by developing a dynamic k-means variant known as X-means that performs bisecting splits of clusters where the resulting clustering solution is addressed by the Bayesian Information Criterion (BIC). Here the bisecting k-means algorithm splits the data samples into two disjoint clusters, and if the split increases the overall fitness measured by internal validity indices, the cluster is split, and the bisecting k-means continues recursively. If none of the bisecting steps of each existing cluster leads to an improved result, the algorithm stops and returns the current clusters as a result. In Muhr and Granitzer [20] a comparison is proposed between a X-means method with several internal multivariate validity indices that measure cluster quality based on the total scatter of between-cluster and within-cluster sum of squares, as the Calinski and Harabasz index (see Caliński and Harabasz 4), the Hartingan index (see Hartigan 9) and the Krzanowski and Lai index (see Krzanowski and Lai 15).

The choice of the metric is another key argument that has been analyzed before among the preliminary choices of a clustering approach. With the most common Euclidean distance, the k-means algorithm tends to detect spherical clusters, and this may not be always a good choice a priori. A variant is the density-sensitive

k-means algorithm that is able to identify non-convex clustering structures, thus generalizing the application area of the original k-means algorithm.

Concerning the clustering of time series, we know that these kind of algorithms work better with time series of equal length because the concept of cluster centers becomes unclear in case of clusters containing series of unequal length.

1.6 Neural Network Models

Artificial neural networks are motivated by biological neural networks and have been used extensively for both classification and clustering purposes.

A neural network model consists of a set of cells that are computational units and a set of one-way data connections joining units as we can see in Fig. 1.2.

Units receive their inputs and compute as outputs several activations, corresponding to signed numbers. The activations reach other units through the connections which are characterized by weight determining whether an activation influences the receiving cell to produce a similar or a different activation and the magnitude of the influence.

The main appeal of neural network is the ability of the systems to automatically learn and improve from experience without being explicitly programmed using complex data inputs. We talk about machine learning when we consider computer models improving their performance based on their input data. We classify machine learning into *supervised* and *unsupervised* models. Supervised learning involves supplying additional input data that measures the performance of a program during a training phase. Supervised learning is the machine learning task of learning a function that maps an input to an output based on example input-output pairs. It infers a function from labelled training data consisting of a set of training examples. On the other hand, unsupervised learning has no performance evaluation tool and its main goal is to construct groups of similar input patterns: they represent the clustering models. The input to these learning algorithms is a set of training examples $\{E^k\}$ that are simply vectors representing patterns or a useful interpretation of them, each one associated to an input cell; for supervised learning

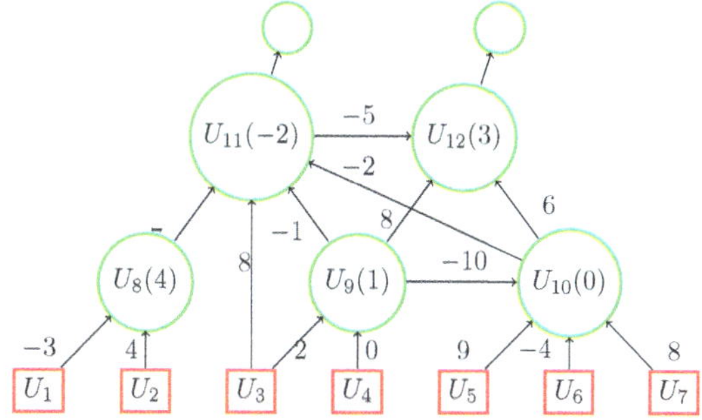

Fig. 1.2 Connectionist model showing weights for connections. Squared cells stand for the input cells, while the round cells represent the intermediate/output cells. The numbers in brackets inside cells are biases

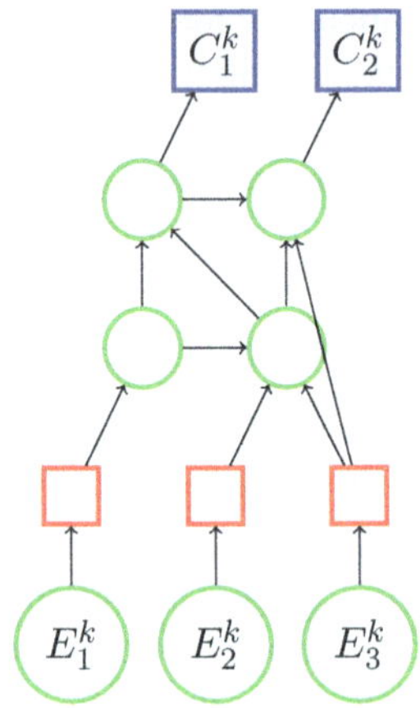

Fig. 1.3 Training example inputs and outputs for a supervised learning model. Training examples are noted as E^k, network output desired responses as C^k, and red squared nodes stand for the input cells

each example is also associated with a right response C^k from the network's output cells and we talk about training example to encompass the −tuple (E^k, C^k). A supervised learning model is represented in Fig. 1.3.

On the other hand, the unsupervised learning problems does not specify the right response for the network, i.e., there is no feedback to correctly classify, and we only see the uncategorized data $\{E^k\}$.

Learning algorithms are classified as one-shot or iterative, depending whether the training examples are taken into consideration only once or several times. In this latter instance, the cycling goes either through the set of the given examples, or it may be determined by choosing one example at random for every cycle in the course of the computation of a set of weights for the given network. Iterative algorithms are more powerful than one-shot ones, but in general they tend to be slower.

A second main appeal of a neural network model has to do with the way information is recorded and organized. For neural network models, knowledge representation consists of network, connection weights, and semantic interpretations attached to cells and activations. Similar patterns are grouped by the network and represented by a single unit called neuron. Specifically, the network is composed by several neurons that are joined by direct connections. A connection has a weight $w_{i,j}$, specifying the influence of neuron u_j on neuron u_i; a positive weight means reinforcement, while a negative one represents inhibition. The set of cells $\{u_1, \ldots, u_p\}$ representing the externally given inputs of the model are called network inputs, while the network outputs, composed by output cells, is the final product; intermediate cells are not seen externally and are called hidden units. The hidden units are functional on generating network outputs through a working process that could be *recurrent*, if it contains several cycles or *feed-forward* if it does not contain them.

The cells of a network are frequently organized into layers $L_0, \ldots, L_k$ such that if $u_j \in L_u$ is connected to $u_i \in L_v$ then $u < v$. If in addition all cells are connected only to cells in the next layer, i.e., $v = u + 1$, we talk about *strictly layer networks*.

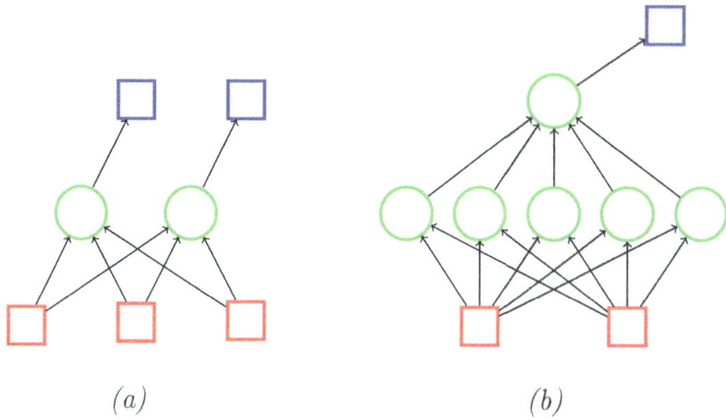

Fig. 1.4 Single-layer network (**a**) and flat network (**b**)

Two important classes of networks are *single-layer* and *flat* networks which are represented in Fig. 1.4; the first ones are strictly one-layer networks where every input cell is connected to every output cell, while flat networks are strictly two-layer where all input cells are in L_0, all output cells are in L_2, and every cell in L_u is connected to every cell in L_{u+1} for $u = 0, 1$.

Let's have a look at the cell properties. We said before that each cell u_i is connected to other cells and its activation must be computed from the activations of cells directly connected to it and the corresponding weights. The new activation u_i is a weighted sum of the inputs to cell u_i from directly connected cells, $u_i = f(S_i)$ where $f(.)$ is the activation function and $S_i = \sum_{j=0}^{n} w_{i,j} u_j$.

Given the dependence on the weighted sum, this type of cell is called *semi-linear cell*; this classification is independent of the activation function that could be deterministic (it is usually a nonlinear, bounded and piece-wise differentiable function) or stochastic. Usually in feed-forward models from a dynamical point of view, cells are visited in a fixed order, each cell reevaluating and changing its activation before the next one is visited; this way steady state is achieved after one pass. On the other hand, for recurrent models we can make the same ordered pass through the cells, or alternatively we can make repeated passes through the network or compute new activations for all cells simultaneously; anyway all these possibilities do not guarantee to reach the steady state.

1.6.1 Clustering Algorithms

Although unsupervised learning doesn't use a training set to provide correct classifications, there must be some criterion for a clustering decisions algorithm and then they have in fact a fixed, built-in teacher. In the following we will look at Kohonen's topology-preserving maps which use the topological arrangement of the learning cells to help capture information.

1.6.2 Kohonen Self-Organizing Maps

Teuvo Kohonen Kohonen [12] has been the primary developer of topology-preserving maps; the main idea is to consider the output cells as occupying positions in space and to take into consideration the *neighbors* of a particular cell. The neighborhood of a cell consists of those cells whose distance from the first one is less than or equal to a fixed threshold. We would like the behavior of the cells to be consistent with the neighborhood structure so that similar inputs lead to output cells being activated that are close together.

We assume there is a $n \times n$ array of output cells, p input cells, and a set of training examples $\{E^k\}$ that correspond to the input cells. The first goal is to associate the inputs by the output cells, i.e., for every training example E^k, we select at least one output cell to have weights close to E^k in a chosen distance metric as much as possible. The second goal is to have close output cells which behave similarly because they have similar weighted sums; this yield to a topology-preserving map.

The mapping from a higher dimensional input space to a lower dimensional output space, through the previous procedure, is called (Kohonen) self-organizing map (SOM in the following). More formally, Kohonen map is a two-layer, feed-forward network such that every output cell has incoming connection from every input cell. For each output cell u_j, the weights of the corresponding input layer connections form a vector of weights $\mathbf{w}_j$ of dimension equal to the number of units in the input layer, p, which can therefore be plotted in the same space of the input data. Output units compete so that, given an input cell $\mathbf{x}$, only one input cell $u_j(\mathbf{x})$ will become activated. This competition is based on the similarity between the input vector and each weight vector measured on a particular distance, for example, the Euclidean distance $\|\mathbf{w}_j - \mathbf{x}\|$. Thus, the active output cell is $u_j(x)$ such that

$$j = \arg \min_{j \in \{1,\ldots,p\}} \|\mathbf{w}_j - \mathbf{x}\|.$$

In the learning phase, the weight vector of the active output is updated so that it becomes closer to the training example by an amount proportional to a learning rate α. Even if in the conventional competitive learning model, only the winning cell should update its weight, lateral interactions between the output cells are required to develop the self-organization. Kohonen proposed to modulate weight adaption according to a Gaussian neighborhood function $\Lambda : (j, r) \in \{1, \ldots, p\}^2 \rightarrow [0, 1]$, such that

$$\Lambda(u_j, u_j(\mathbf{x})) = \exp\left(-\frac{\|u_j - u_j(\mathbf{x})\|^2}{2\sigma^2}\right),$$

where $\|u_j - u_j(\mathbf{x})\|$ is the distance between any output cell u_j and the activated one $u_j(\mathbf{x})$ given a certain metric defined over the input layer and where σ is the size of the neighborhood around an output cell, i.e., how far should an output cell be from the activated one so as not to be affected by it. The learning rule is

$$\mathbf{w}_y \leftarrow \mathbf{w}_j + \alpha \Lambda(u_j, u_j(\mathbf{x}))(\mathbf{x} - \mathbf{w}_j).$$

As said before, the output layer metric induces a topology. Provided a good match between the output layer topology and the topology of the input data exists, the network will develop into a topology-preserving map: the best matching output units $u_j(\mathbf{x}_1)$ and $u_j(\mathbf{x}_2)$ corresponding to two input cell vectors $\mathbf{x}_1$ and $\mathbf{x}_2$, that are close in the input space, will be also close in the output layer metric.

Kohonen [13] suggested shrinking the neighborhood size as the algorithm proceeds, in order to accelerate the convergence of the expected squared quantization error to a local minimum. In the Gaussian choice, that is done by slowly decreasing its width $\sigma(t) \to 0$ as $t \to \infty$.

The system is said to be stable if no pattern in the training data changes its category after a finite number of iterations. To assure the stability of the system, the learning rate should be decreased to zero as iterations progress. This problem is linked to the ability of the algorithm to adapt to new data, called plasticity.

Like the previously discussed methods, i.e., k-means and fuzzy k-means algorithms, SOM does not work well with time series of unequal length due to the difficulty involved in defining the dimension of weight vectors.

1.7 Search-Based Approaches

Search techniques are used to recover the optimum value of a criterion function. We distinguish between deterministic and stochastic search techniques; the deterministic ones guarantee an optimal partition by performing exhaustive enumeration, while the stochastic ones generate a near-optimal partition quickly and converge to optimal partition asymptotically. We classify all the models analyzed before as deterministic search techniques. In the following we describe two stochastic search techniques.

1.7.1 Evolutionary Approaches for Clustering

Evolutionary approaches motivated by natural evolution are among the stochastic search techniques. They suggest to apply evolutionary operators like selection, recombination, and mutation, to a population of solutions to obtain the globally optimal partition of data. Evolutionary operator transforms input chromosomes into an output chromosomes that are selected as survivors into the next generation by a fitness function.

The best-known evolutionary techniques are genetic algorithms proposed for the first time in [24] to minimize the squared error of a clustering. Here each point or chromosome represents a partition of n objects into k clusters, and the optimal partition is recovered through a genetic algorithm. A number of papers approach the optimization problem posed by clustering using genetic algorithms as the optimization tool. These algorithms have long been used in different kinds of complex problems, usually with encouraging results. Sometimes a genetic algorithm is used to optimize the objective function used in different approaches as the k-means algorithm, where this measure of within cluster variance is used as a fitness function. In case of low-quality data especially if they are multidimensional which varies in precision and resolution, genetic algorithm should be capable of providing satisfactory classifications and tackle the challenges prompted by the presence of some fuzziness in data. In this case genetic algorithm performs better than the traditional methods like k-means.

1.7.2 Simulated Annealing Approach

The simulated annealing approach is a sequential stochastic search technique whose applicability to clustering is discussed in [11]. The algorithm requests to select values for the control parameters governing the probability of acceptance, called initial and final temperatures, T_0 and T_f, respectively. The algorithm is based on the following steps:

i Randomly select an initial partition P_0 and compute the squared error value E_{P_0}.
ii Select a neighbor P_1 of P_0 and compute its squared error E_{P_1} and compare it with E_{P_0}: if $E_{P_1} > E_{P_0}$ assign P_1 to P_0 with a temperature-dependent probability else, assign P_1 to P_0. Repeat for a fixed number of iterations.
iii Reduce the initial temperature as $T_0 = cT_0$ for a constant c, and compare it with T_f: if $T_0 > T_f$ repeat from [ii] else stop.

Simulated annealing guarantees to reach the global optimal solution. Therefore it can be slow in reaching the optimal solution because optimal results require the temperature to be decreased very slow from iteration to iteration.

Many combinations of several algorithms are proposed in literature; [18] proposed an optimized clustering algorithm based on SOM which employs the rough set theory and the simulated annealing technique as a general algorithm for optimization problems (see [21, 22]). The goal of simulated annealing technique is to find a global optimum, in terms of some clustering error, in the presence of a local optimum by shaking the cells vigorously for a time then shaking it slightly less vigorously for another period and so on until we barely shake it at all; the basic idea is that during one of the shaking periods, there will be just enough "energy" to move an output cell from one group to another leading to a minor clustering

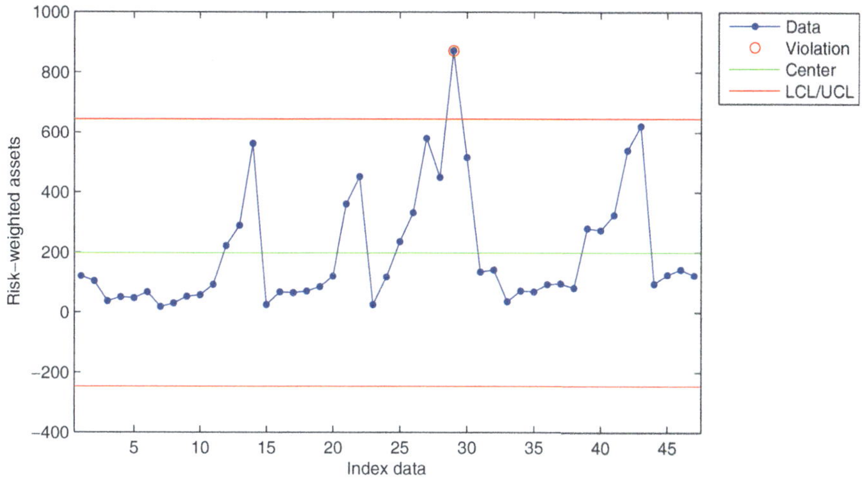

Fig. 1.5 Risk-weighted assets data

error.[2] Here the optimal number of clusters can be found by rough set theory which groups the given cells into overlapping clusters. Then the overlapped cells will be assigned to the true clusters they belong to, by applying simulated annealing algorithm which optimizes the clustering operation by assigning the nearest cluster center to the overlapped cells and then minimizes the error that comes from the clustering operations in order to be as precise as possible.

1.8 A Clustering Exercise on the European Banking System

For a comparison of three methodologies discussed above, we present a clustering exercise on a sample of the major 47 European banks. The sample is taken from [2] to which we added Dexia and KBC. The sample collects the data from major European banks that were subject to the stress test exercise by the European regulators in September 2010. For the clustering, we used three variables: (i) risk-weighted assets, (ii) leverage, and (iii) tier I capital. In Figs. 1.5, 1.6, and 1.7, we represent the risk-weighted assets, the leverage and the tier 1 risk data, respectively, plotted with evidence of the center line at the data average and the upper/lower control limits at three standard errors from the center line. We observe that data are quite regular because we record only one violation in risk-weighted asset and in leverage ratio data. In Tables 1.1, 1.2, and 1.3, we report the cardinalities of the

[2]This kind of model is very familiar to physicists, and they represent the vigor of the shaking as a temperature because at higher temperature, there is more molecular activity. Annealing refers to a way of tempering certain alloys of metal by heating and then gradually cooling them.

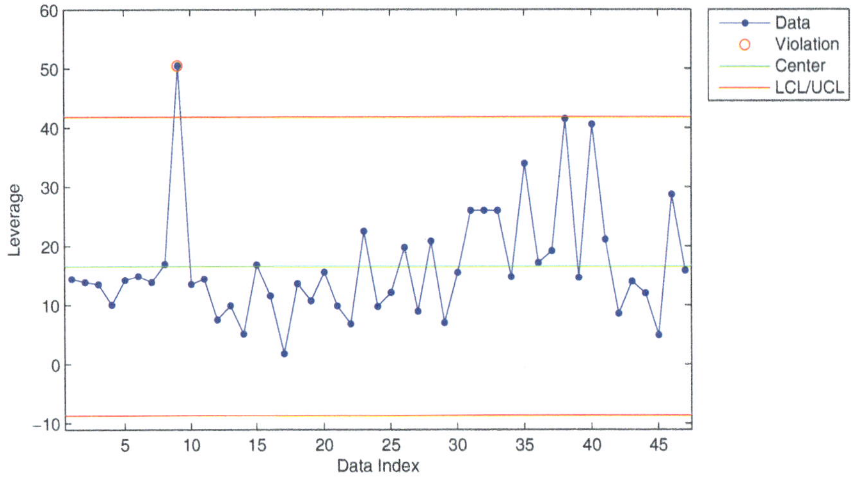

Fig. 1.6 Leverage data

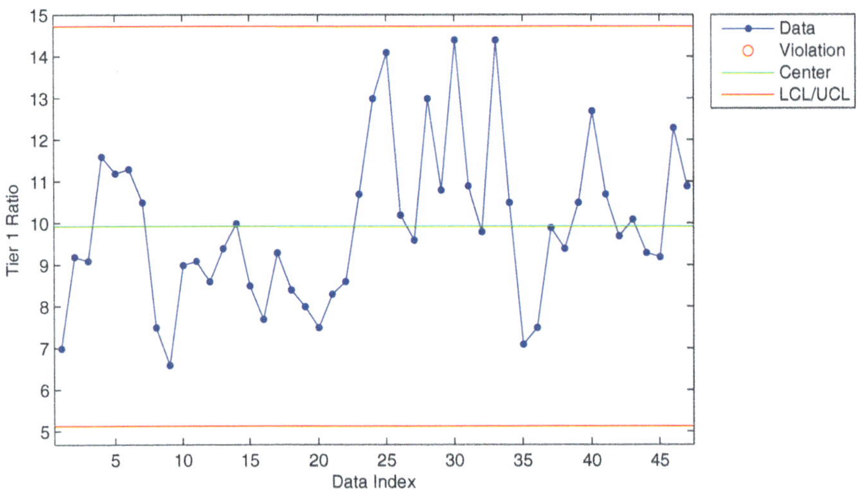

Fig. 1.7 Tier 1 risk data

clusters used with three methods (k-Means, SOM, and hierarchical clustering) and the centroids recovered considering 8 clusters.

To analyze the impact of the chosen number of clusters in the comparison of the clustering methods, we run k-means, SOM, and the hierarchical procedure choosing a number of cluster also equal to 9. We point out that implementing SOM procedure with a network of dimension 3×3, anyway data result grouped in eight clusters since a neuron remains empty. In Figs. 1.8 and 1.9, we compare the clusters' cardinalities generated by the procedures choosing different number of clusters. We observe that

Table 1.1 *k*-means Clusters' centroids for the three clustering data compared with the sample's means

... K-means	Risk-weighted asset	Leverage	Tier 1 ratio
Cluster 1	574.07	10.95	10.77
Cluster 2	290.16	17.04	10.56
Cluster 3	66.89	11.79	9.37
Cluster 4	94.16	14.66	8.83
Cluster 5	67.51	42.09	7.7
Cluster 6	128.91	17.76	10.07
Cluster 7	27.39	19.34	10.32
Cluster 8	45.37	12.69	10.63
Sample means	199.7	16.7	9.9

Table 1.2 SOM Clusters' centroids for the three clustering data compared with the sample's means

... SOM	Risk-weighted asset	Leverage	Tier 1 ratio
Cluster 1	260.61	17.05	11.06
Cluster 2	67.09	21.89	8.81
Cluster 3	531.55	11.49	10.77
Cluster 4	128.91	17.76	10.07
Cluster 5	34.13	16.84	10.43
Cluster 6	871.69	7.1	10.08
Cluster 7	339.4	17.01	9.73
Cluster 8	94.16	14.66	8.83
Sample means	199.7	16.7	9.9

Table 1.3 Hierarchical clusters' centroids for the three clustering data compared with the sample's means

... H	Risk-weighted asset	Leverage	Tier 1 ratio
Cluster 1	281.22	21.8	10.87
Cluster 2	339.4	17.01	9.73
Cluster 3	620.71	14.09	10.1
Cluster 4	549.4	9.64	10.92
Cluster 5	229.69	9.91	11.35
Cluster 6	451.28	13.92	10.8
Cluster 7	79.78	18.13	9.56
Cluster 8	871.69	7.1	10.81
Sample means	199.7	16.7	9.9

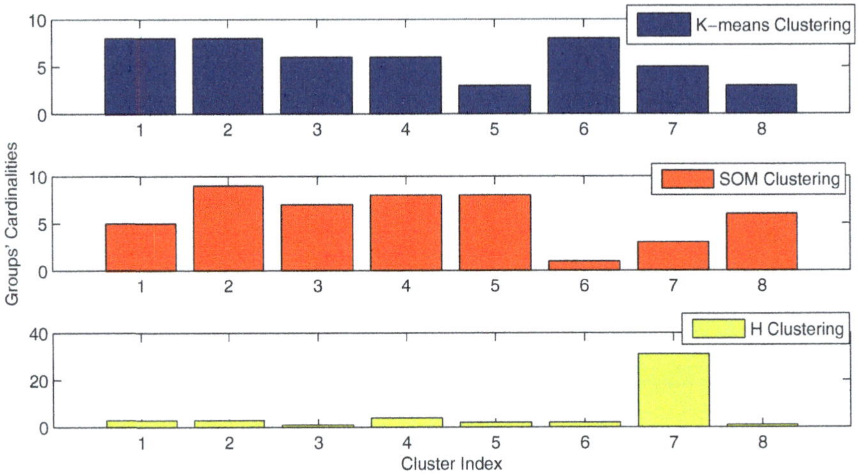

Fig. 1.8 Comparison between the clusters' cardinalities generated by SOM approach and k-means and hierarchical imposing a number of clusters equal to 8

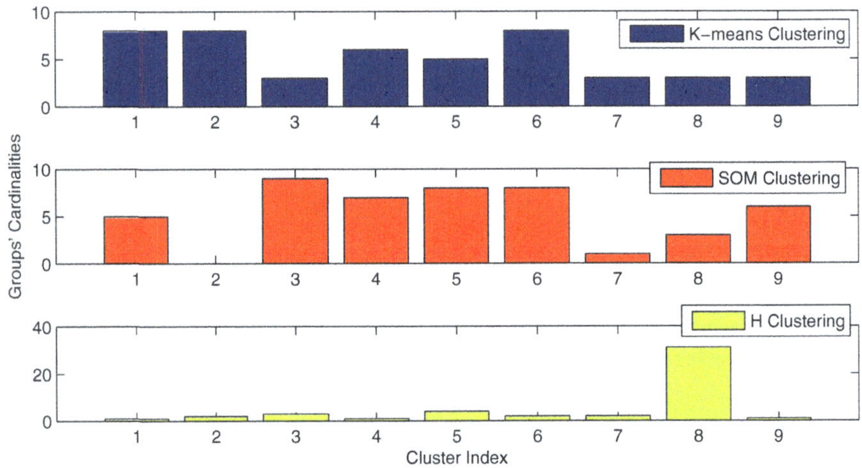

Fig. 1.9 Comparison between the clusters' cardinalities generated by SOM approach and k-means and hierarchical imposing a number of clusters equal to 9

hierarchical clustering is very different of the other methods and the final result is to concentrate a high number of data into one single cluster.

In Fig. 1.10 we compare the centroids' levels of data computed with several clustering methods, while in Fig. 1.11 we compare single data with respect to the cluster assigned by the clustering approaches, with a number of cluster equal to 8.

Hierarchical clustering allows to generate a hierarchical tree represented in Fig. 1.12a corresponding to a cophenetic correlation coefficient equal to 0.8489. The

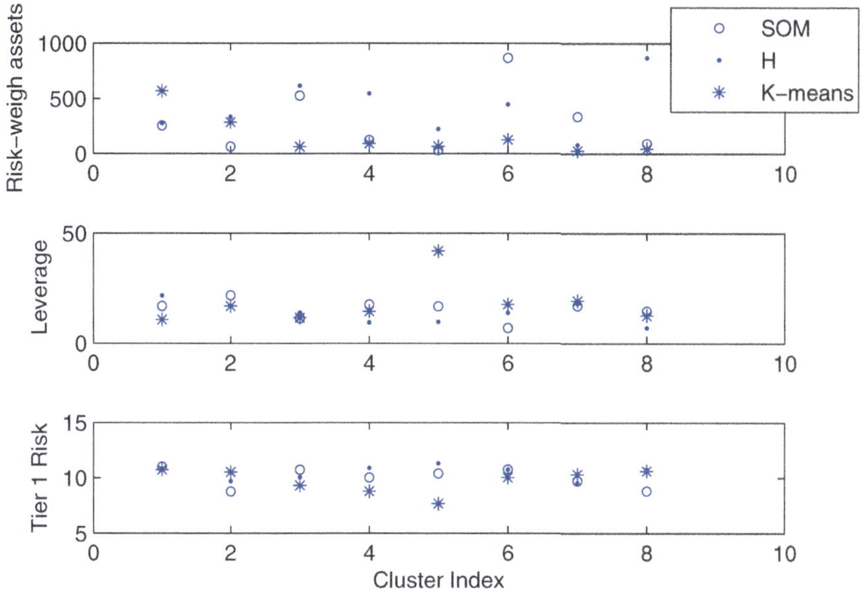

Fig. 1.10 Comparison between the clusters' centroids generated by k-means, SOM, and hierarchical approach with a number of clustering equal to 8

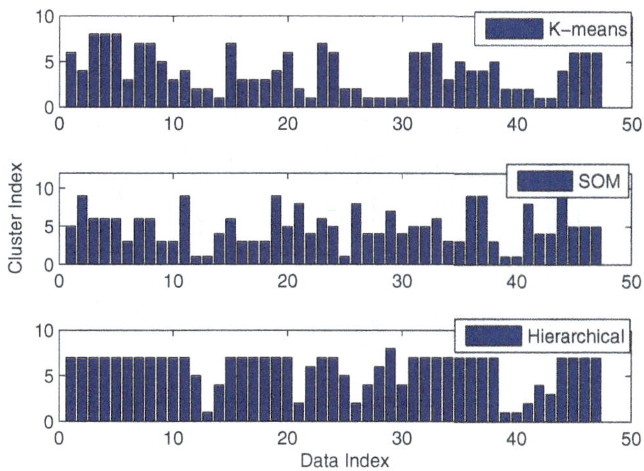

Fig. 1.11 Comparison between the clusters assigned to each data by k-means, SOM, and hierarchical approach with a number of clustering equal to 8

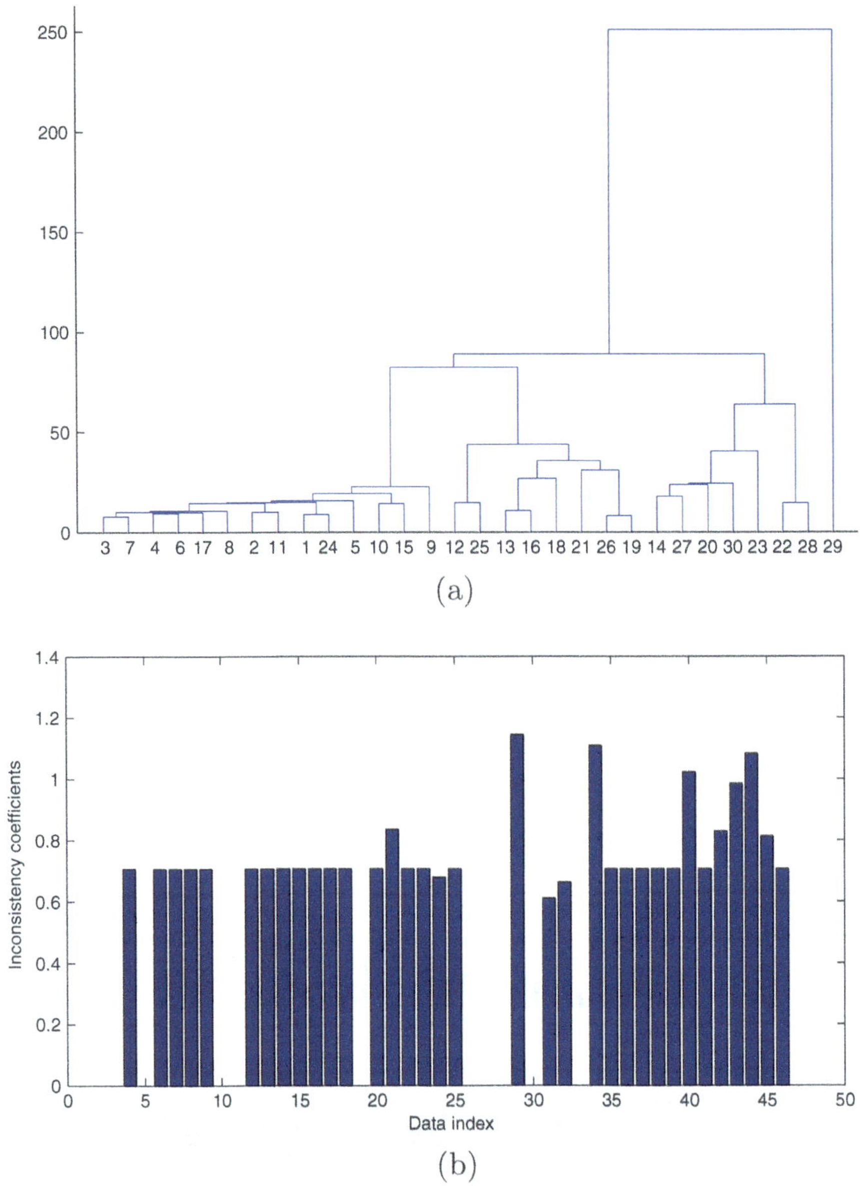

Fig. 1.12 Hierarchical tree: dendrogram and inconsistency coefficients. (**a**) Dendrogram of the Hierarchical tree of data. (**b**) Inconsistency coefficients of data

dendrogram shows inconsistent links. Note how the objects in the dendrogram fall into two groups that are connected by links at a much higher level in the tree. These links are inconsistent when compared with the links below them in the hierarchy.

The relative consistency of each link in a hierarchical cluster tree can be quantified and expressed as the inconsistency coefficient. This value compares the height of a link in a cluster hierarchy with the average height of links below it. Links that join distinct clusters have a high inconsistency coefficient; links that join indistinct clusters have a low inconsistency coefficient. The hierarchical clustering is determined by computing the inconsistency coefficients of data represented in Fig. 1.12b.

References

1. Anderberg, M. R. (1973). *Cluster Analysis for Applications*. New York: Academic.
2. Baglioni, A., & Cherubini, U. (2010). Marking-to-Market Government Guarantees to Financial Systems: An Empirical Analysis of Europe, working paper available at SSRN: https://doi.org/10.2139/ssrn.1715405.
3. Bernardi, E., & Romagnoli, S. (2011). Computing the volume of an high-dimensional semi-unsupervised hierarchical copula. *International Journal of Computer Mathematics, 88*(12), 2591–2607.
4. Caliński, R., & Harabasz, J. (1974). A dendrite method for cluster analysis. *Communications in Statistics, 3*, 1–27.
5. Dunn, J. C. (1974). Well separated clusters and fuzzy partitions. *Journal on Cybernetics, 4*, 95–104.
6. Francois, D., Wertz, V., & Verleysen, M. (2005). Non-Euclidean metrics for similarity search in noisy data sets. In *Proceedings of European Symposium on Artificial Neural Networks*, ESANN'2005.
7. Gallant, S. I. (1993). *Neural Network Learning and Expert Systems*. Cambridge: MIT Press.
8. Gowda, K. C., & Krishna, G. (1977). Agglomerative clustering using the concept of mutual nearest neighborhood. *Pattern Recognition, 10*, 105–112.
9. Hartigan, J. (1985). Statistical theory in clustering. *Journal of Classification, 2*(1), 63–76.
10. Jain, A. K., Murty, M. N., & Flynn, P. J. (1999). Data clustering: A review. *ACM Computing Surveys, 31*(3), 264–323.
11. Klein, R. W., & Dubes, R. C. (1989). Experiments in projection and clustering by simulated annealing. *Pattern Recognition, 22*(2), 213–220.
12. Kohonen, T. (1982). Self-organized formation of topologically correct feature maps. *Biological Cybernetics, 43*, 59–69.
13. Kohonen, T. (1991). Self-organizing maps: Optimization approaches. In *Artificial Neural Networks* (pp. 981–990). Amsterdam: Elsevier Science Publisher.
14. Kosmidis, I., & Karlis, D. (2014). Model-based clustering using copulas with applications. *Statistics and Computing*, Source: arXiv:1404.4077v5.
15. Krzanowski, W. J., & Lai, Y. T. (1985). A criterion for determining the number of groups in a data set using sum of squared clustering. *Biometrics, 44*, 23–34.
16. Liao, T. W. (2005). Clustering of time useries data-a survey. *Pattern Recognition, 38*, 1857–1874.
17. MacQueen, J. B. (1967). Some methods for classification and analysis of multivariate observations. In *Proceedings of the 5th Symposium on Mathematical Statistics and Probability* (pp. 281–297). Berkeley: University of California Press.

18. Mohebi, E., & Sap, M. N. M. (2009). Hybrid Kohonen self organizing map for the uncertainty involved in overlapping clusters using simulated annealing. In *Proceedings of the UKSim 2009: 11th International Conference on Computer Modelling and Simulation* (pp. 53–58).
19. Montero, P., & Vilar, J. A. (2014). TSclust: An R package for time series clustering. *Journal of Statistical Software, 62*(1), 1–43.
20. Muhr, M., & Granitzer, M. (2009). Automatic cluster number selection using a split and merge K-means approach. DEXA Workshops 2009 (pp. 363–367).
21. Munakata, T. (2008). *Fundamentals of the New Artificial Intelligence*. New York: Springer.
22. Pawlak, Z. (1982). Rough sets. *International Journal of Computer and Information Science, 11*, 341–356.
23. Pelleg, D., & Moore, A. (2000). X-means: Extending K-means with efficient estimation of the number of clusters. In ICML-2000 (pp. 727–734).
24. Raghavan, V. V., & Birchand, K. (1979). A clustering strategy based on a formalism of the reproductive process in a natural system. In *Proceedings of the Second Annual International ACM SIGIR Conference on Information Storage and Retrieval, Information Implications into the Eighties*, Dallas, 27–28 Sept 1979 (pp. 10–22). New York: ACM Press.
25. Su, M. C., & Chou, C. H. (2001). A modified version of the K-means algorithm with a a distance based on cluster symmetry. *IEEE Transactions on Pattern Analysis and Machine Intelligence, 23*(6), 674–680.
26. Wang, L., Bo, L., & Jiao, L. (2006). A modified K-means clustering with a density-sensitive distance metric. In *Rough Sets and Knowledge Technology* (Lecture notes in computer science, Vol. 4062, pp. 544–551). Berlin/Heidelberg: Springer.
27. Wilson, D. R., & Martinez, T. R. (1997). Improved heterogeneous distance functions. *Journal of Artificial Intelligence Research, 6*, 1–34.
28. Wishart, D. (1969). An algorithm for hierarchical classifications. *Biometrics, 25*(1), 165–170.

Chapter 2
Copula Function and C-Volume

Abstract This chapter is devoted to a short review of the basic concepts of the theory of dependence functions or copula functions, as they are more usually called. In particular our objective is to define the volume of a copula function: this is one of the basic ingredients of the aggregation algorithm that will be discussed in Part II.

After a review of the fundamentals of copula theory, we will address our attention to explaining the meaning of the volume of a copula function in view of its pivotal role in the aggregation algorithm that will be discussed in the following section. The classical books by Nelsen [25] and Joe [20] are recommended readings for those willing to further investigate this subject matter.

2.1 Copula Functions

Copula function is a useful representation of a n-dimensional joint distribution function which enables one to separate the specification of marginal distributions from the dependence structure. The key result upon which all of this theory is built is the theorem of *probability integral transformation* where it is proven that if one takes a random variable X with continuous distribution $F_X(x)$, $\forall x \in \mathbb{R}$, the variable $u \equiv F_X(x)$, $\forall x \in D_X$, where D_X denotes the domain of X, has the uniform distribution in the unit interval. This result may help to rewrite a joint distribution function of $(X_1, \cdots, X_n)$ in an interesting way. In the n-dimensional case, given $X_i, i = 1, \ldots, n$ with continuous distributions $F_{X_i}(x_i), x_i \in \mathbb{R}, i = 1, \ldots, n$, whose generalized inverse functions[1] are $F_{X_i}^{[-1]}(u_i) = x_i, u_i \in [0, 1], x_i \in \mathbb{R}, i = 1, \ldots, n$, we rewrite the joint distribution in terms of uniformly distributed variables as

$$F_{X_1,\ldots,X_n}(x_1, .., x_n) = F(F_{X_1}^{-1}(u_1), \ldots, F_{X_n}^{-1}(u_n)) \equiv C(\mathbf{u}),$$

[1] We define the *generalized inverse* $F^{[-1]} : (0, 1) \to \mathbb{R}$ as $F^{[-1]}(w) = \inf\{l \in \mathbb{R} : F(l) \geq w\}$, that coincides with the standard inverse function when this exists.

E. Bernardi, S. Romagnoli, *Counting Statistics for Dependent Random Events*,
https://doi.org/10.1007/978-3-030-64250-1_2

where $u_i = F_{X_i}(x_i)$. These uniform variables are called *marginal distributions*, while the function that links them is called the *copula function*.

The requirements that have to be met by the function $C(\mathbf{u})$ in order to have a well-defined joint distribution are summarized in the following definition.

Definition 2.1.1 A copula is a function $C : [0, 1]^n \rightarrow [0, 1]$ satisfying the following requirements

1. *Grounded*: $C(u_1, \ldots, u_{i-1}, 0, u_{i+1}, \ldots, u_n) = 0, \forall i \in \mathbb{N} \subset (1, n)$.
2. *Copula marginal*: $C(u_1, \ldots, u_{i-1}, 1, u_{i+1}, \ldots, u_n), \forall i \in \mathbb{N}$ is a $(n-1)$-copula function.
3. *n-Increasing*: $V_C(\mathbf{S}) \geq 0, \forall \mathbf{S}$ where $\mathbf{S}$ is a n-dimensional box defined by any couple of vectors $[\mathbf{u}, \mathbf{v}]$.

The first item in the Definition assures that the joint probability of n events is zero if at least one of these has a zero marginal probability to happen. On the other hand, if one event will take place for sure, the joint probability will be equal to the joint probability of other events. The last requirement involves the concept of the *volume* of a n-dimensional figure representing the joint probability that variables will be in some given box. It also refers to a non-negativity condition on that volume.

The relationship between copula function and joint distributions is clearly explained in *Sklar's theorem*, which is the basic pillar upon which the theory of copula functions has been built. *Sklar's theorem* claims that not only copula functions are joint distributions functions, but the converse is also true, i.e., every joint distribution function can be written as a copula function taking the *marginal* distribution functions as arguments. This way copulas arise as the natural tool for statistical dependence modeling through *Sklar's theorem*.

Theorem 2.1.1 (Sklar [31]) *Let $(X_1, \ldots X_n) \equiv F_{X_1,\ldots,X_n}$, where $X_i \sim F_{X_i}, i = 1, \ldots, n$. There exists a n-dimensional copula C such that*

$$F_{X_1,\ldots,X_n}(x_1, .., x_n) = C(F_{X_1}(x_1), \ldots, F_{X_n}(x_n)),$$

where $x_i \in \mathbb{R}, i = 1, \ldots, n$. Moreover if $F_{X_i}, i = 1, \ldots, n$ are continuous, then C is unique. Conversely, if C is a n-dimensional copula and $F_{X_i}, i = 1, \ldots, n$ are distribution functions, then $F_{X_1,\ldots,X_n}$ is a n-dimensional distribution function.

Proof See Nelsen [25]. □

Sklar's theorem hence highlights the link between multivariate distributions functions and their univariate marginals providing a representation where we can separate the specification of the joint distribution from that of marginal distributions. We observe that if the joint distribution function is absolutely continuous and the marginals are strictly increasing, we are able to write the joint density in term of the marginals' densities and the density of the copula functions, i.e.,

$$f_{X_1,\ldots,X_n}(x_1, .., x_n) = c(F_{X_1}(x_1), \ldots, F_{X_n}(x_n)) \prod_{i=1}^{n} f_{X_i}(x_i),$$

where $x_i \in \mathbb{R}, i = 1, \ldots, n$ and the densities are noted in small letters.

2.1.1 Fréchet-Hoeffding Bounds of a n-Dimensional Copula and Association Measures

It is important to analyze copula functions defining the range of the possible dependence structure and their extreme values.

The following theorem provides the extreme values of a copula function.

Theorem 2.1.2 *Every copula satisfies the following inequality:*

$$C^{-}(\mathbf{u}) \leq C(\mathbf{u}) \leq C^{+}(\mathbf{u}), \forall \mathbf{u} \in [0, 1]^n$$

where the minimal bound $C^{-}(\mathbf{u})$ and the maximal one $C^{+}(\mathbf{u})$ are defined as

$$C^{-}(\mathbf{u}) = \max(\sum_{i=1}^{n} u_i - 1, 0)$$

$$C^{+}(\mathbf{u}) = \min(\mathbf{u}), \forall \mathbf{u} \in [0, 1]^n,$$

where $\mathbf{u}$ is the vector of marginals, i.e., $u_i = F_{X_i}(x_i), i = 1, \ldots, n$.

Proof See Nelsen [25]. □

The set $\{C^{-}, C^{+}\}$ defines the Fréchet-Hoeffding bounds of a n-dimensional copula. The upper Fréchet-Hoeffding bound, which always satisfies the definition of copula function, is called *maximal copula*. However the lower bound never satisfies the necessary conditions to be defined as a copula function for $n > 2$ (see Féron 15). It is important to observe that in the two-dimensional case, Fréchet-Hoeffding bounds represent perfectly negatively and positively dependent structures, respectively, suggesting that the value of copula functions increase with the dependence. However in a n-dimensional setting, concordance order between copulas is more involved, and it is not possible to interpret Fréchet-Hoeffding bounds as perfectly dependence structures. Nevertheless the case of no dependence at all is notably and it is represented by the *product copula* defined as

$$C^{\top}(\mathbf{u}) = \prod_{i=1}^{n} u_i.$$

Figure 2.1 reports the scatter plots of the Fréchet-Hoeffding bounds along with the product copula, allowing to understand their meaning in term of perfect dependence and independence, respectively.

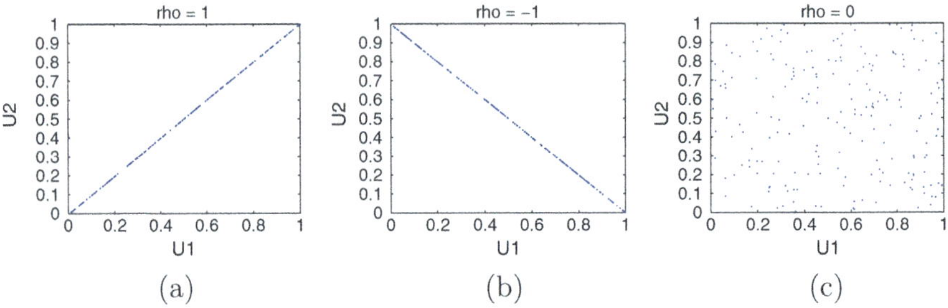

Fig. 2.1 Scatter plots of 200 data simulated from the bivariate Fréchet-Hoeffding bounds copulas and the product copula. (**a**) Minimal copula. (**b**) Maximal copula. (**c**) Product copula

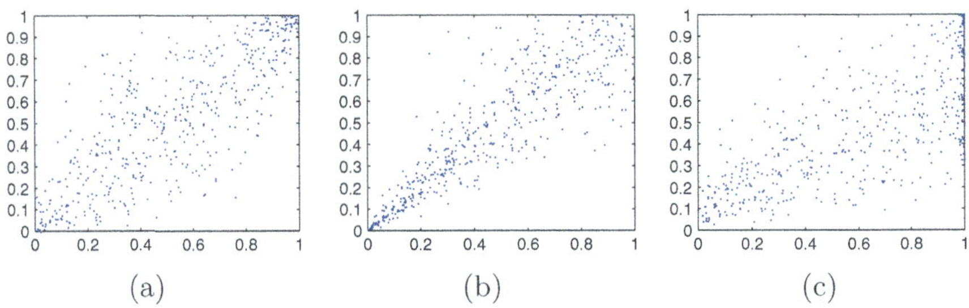

Fig. 2.2 Scatter plots of 500 data simulated from a bivariate radially symmetric and exchangeable copula, from a radially asymmetric and exchangeable copula and from a radially asymmetric and not exchangeable copula. (**a**) Radially symmetric and exchangeable copula. (**b**) Radially asymmetric and exchangeable copula. (**c**) Radially asymmetric and not exchangeable copula

Copulas can also be characterized with respect to different notions of symmetry. We consider in the following two main kinds of symmetry.

Definition 2.1.2 (Radial symmetry) A copula function C is radially symmetric if $C(\mathbf{u}) = C(\mathbf{1} - \mathbf{u})$, where $\mathbf{1}$ denotes the n-dimensional unit vector.

Definition 2.1.3 (Exchangeability) A copula C is exchangeable if $C(\mathbf{u}) = C(\mathbf{u}_\sigma)$, for any entries' permutation $\sigma : 1, \ldots, n \to 1, \ldots, n$.

Figure 2.2 reports the scatter plots of three kind of copula showing different notions of symmetry. We observe that the concept of exchangeability coincides with that of *symmetry* in two-dimensional case, because it holds that $C(u_1, u_2) = C(u_2, u_1), \forall u_i \in [0, 1], i = 1, 2$.

To explain the dependence information concerning two random variables (X_1, X_2) in a single number, it is common to use association measures which are defined in term of their copula function C.

Definition 2.1.4 (Kendall's τ and Spearman's ρ_S) Kendall's τ is given by

$$\tau(C) = 4 \int_{[0,1]^2} C(u_1, u_2) dC(u_1, u_2) - 1,$$

while Spearman's ρ_S by

$$\rho_S(C) = 12 \int_{[0,1]^2} C(u_1, u_2) du_1 du_2 - 3.$$

It holds true that $\tau(C), \rho_S(C) \in [-1, 1]$.

On the other hand to measure the strength of dependence in the tails of the joint distribution it is common to use the tail dependence coefficients, which are also purely copula-based measures.

Definition 2.1.5 (Tail dependence coefficients) The lower tail dependence coefficients is given by

$$\lambda_L(C) = \lim_{t \to 0^+} \frac{C(t,t)}{t},$$

while the upper tail dependence coefficient is given by

$$\lambda_U(C) = 2 - \lim_{t \to 1^-} \frac{1 - C(t,t)}{1-t},$$

provided that the limits exist.[2] It holds true that $\lambda_L(C), \lambda_U(C) \in [0, 1]$.

A copula C is said to be lower/upper tail-dependent if $\lambda_L(C) > 0/\lambda_U(C) > 0$. We observe that a radial symmetric copula C satisfies $\lambda_L(C) = \lambda_U(C)$. It is particularly useful to observe that the characterization of copula function based on the concepts of symmetry and tail dependence can be modified by a rotation of the copula itself. There are different kind of rotations among which we recall the rotation by 180 degrees defining the *survival copula* $\overline{C}$, i.e.,

$$\overline{C}(u_1, u_2) = u_1 + u_2 - 1 + C(1 - u_1, 1 - u_2),$$

where $u_i \in [0, 1], i = 1, 2$. Clearly a radially symmetric copula C coincides with its survival version. Figure 2.3 reports the scatter plots of a copula and its survival.

Finally we observe that the copula function provides a mapping to $[0, 1]$ whose distribution can be characterized using the notion of *Kendall function*.

Definition 2.1.6 (Kendall function) The Kendall function of a copula C is given by

$$K(z, C) = \mathbb{P}(C(\mathbf{u}) \leq z), z \in [0, 1].$$

[2]As for example in the Archimedean family, the assumption to work with strict generators is sufficient to assure the existence of the tail dependence coefficients (see Nelsen 25).

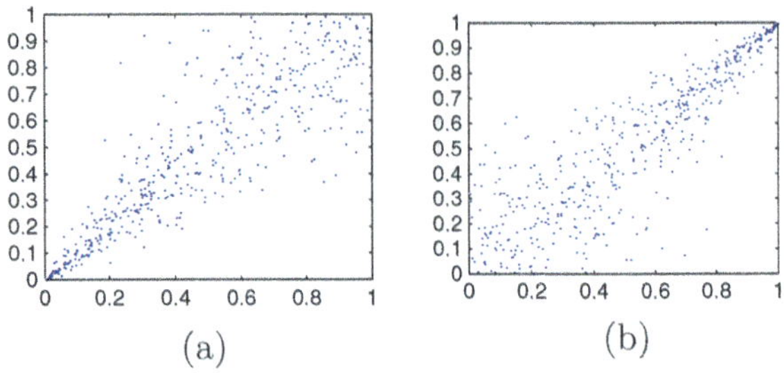

Fig. 2.3 Scatter plots of 500 data simulated from a bivariate copula and its survival. (**a**) No rotation. (**b**) Rotation by 180 degrees

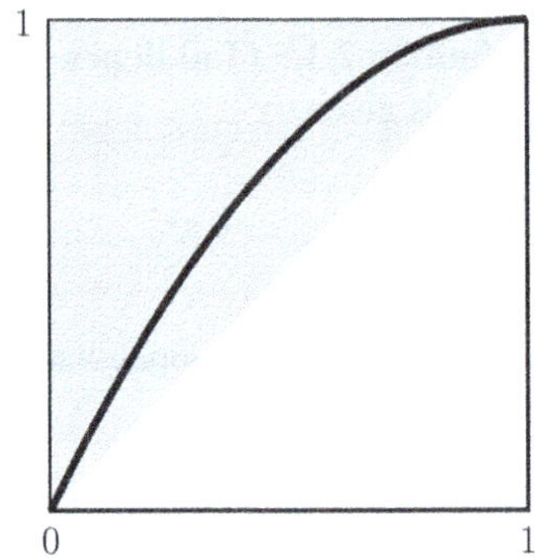

Fig. 2.4 Graphical representation of the Kendall function. The light gray area represents the limiting bounds of the function itself

It holds that $z \leq K(z, C) \leq 1, z \in [0, 1]$, whose bounds represent the case of perfect positive and negative dependence, respectively.

Figure 2.4 gives an example of Kendall function showing its range delimited by the limiting bounds of the function itself.

The computation of the *Kendall function* for a given copula is however complicated in general. Imlahi et al. [19] provide a recursive formula based on the concept of copula quantile function. A convenient exception is the Archimedean copulas that will be introduced in the next section.

2.2 Families of Copulas

We are now going to illustrate the main families of copula functions, namely, the elliptical, the Archimedean, and the extreme-value copulas

2.2.1 Elliptical Copulas

The family of *elliptical copulas* is obtained directly from the class of elliptical distributions, recalling that by *Sklar's theorem* we have

$$C(\mathbf{u}) = F_{X_1,\ldots,X_n}(F_{X_1}^{-1}(u_1), \ldots, F_{X_n}^{-1}(u_n)),$$

where $u_i \in [0, 1], i = 1, \ldots, 2$ and where either the joint distribution or the marginals are elliptical distribution functions. The density of an elliptical copula C is then

$$c(\mathbf{u}) = \frac{f(F_{X_1}^{-1}(u_1), \ldots, F_{X_n}^{-1}(u_n))}{\prod_{i=1}^{n} f_{X_i}(F_{X_I}^{-1}(u_i))},$$

where $u_i \in [0, 1], i = 1, 2$. For the Kendall function of elliptical copulas, however, no close-form expression is available.

It holds true that all elliptical distributions are radially symmetric. For a detailed study on elliptical distribution, we refer the reader to Fang et al. [14]. Here below we propose a discussion about the most famous elliptical distributions being both the Gaussian and the Student's t copulas.

Gaussian Copulas

Define $\Phi_n(\epsilon_1, \ldots, \epsilon_n; \mathbf{R})$ the n-dimension multivariate Gaussian distribution function, linking together a set of n standardized variables with correlation represented by the n-dimension matrix $\mathbf{R}$. It is well-known that the marginal distribution of each variable is also standard normal. Denote $\Phi(x)$ the univariate standard normal distribution function. You can now rewrite the joint normal distribution as

$$C(\mathbf{u}) \equiv \Phi_n\left(\Phi^{-1}(u_1), \ldots, \Phi^{-1}(u_n); \mathbf{R}\right). \tag{2.1}$$

This is known as *Gaussian copula* function, and it is by far the most widely used in financial applications.

The Gaussian copula is defined by $n(n-1)/2$ parameters, unless a specific structure of the correlation matrix is assumed. In the case of an exchangeable correlation matrix with $\rho_{jk} = \rho \in (-1/(n-1), 1)$ for all $j, k = 1, \ldots, n, j \neq k$, the Gaussian copula is exchangeable itself and is specified by only one parameter. Moreover if $\rho_{jk} = 1/0$ for all $j, k = 1, \ldots, n, j \neq k$, we represent the comonotonicity/independent case, while in the bivariate case, we have the countermonotonicity case for $\rho_{12} = -1$.

In the bivariate case, the Gaussian copula depends only on the off-diagonal parameter of the correlation matrix, i.e., $\rho = \rho_{12}$. In this case both the Kendall's τ and Spearman's ρ_S can be represented in term of this parameter as

$$\tau(\rho) = \frac{2}{\pi} \arcsin(\rho)$$

$$\rho_S(\rho) = \frac{6}{\pi} \arcsin\left(\frac{\rho}{2}\right).$$

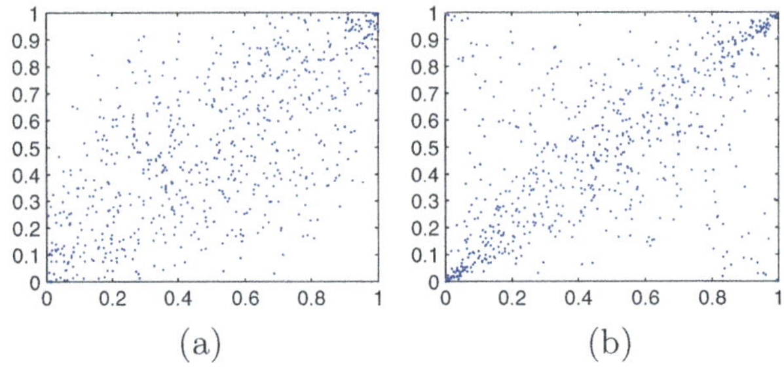

Fig. 2.5 Scatter plots of 500 data simulated from the bivariate Gaussian and t copula with $\rho = 0.7$ and $\nu = 1$. (**a**) Gaussian copula. (**b**) Student's t copula

Finally we observe that the Gaussian copula is tail-independent, i.e., $\lambda_L(\rho) = \lambda_U(\rho) = 0$ thanks to the radial symmetry of the elliptical copulas.

Figure 2.5 reports the scatter plots of the Gaussian and t copula that is presented in the following.

Student's t Copula

We know that in univariate models a way to introduce fat tails is to substitute the normal distribution with the so-called Student's t distribution, in which a parameter ν, representing the *degrees of freedom*, is responsible for generating fat tails. The same can be done with copulas. So, define $\mathbf{T}_n(\epsilon_1, \ldots, \epsilon_n; \mathbf{R}, \nu)$ the joint Student's t distribution function, and denote $\mathbf{T}_\nu(x)$ the univariate Student's t distribution function. The Student's t copula is then defined as

$$C(\mathbf{u}) \equiv \mathbf{T}_n\left(\mathbf{T}_\nu^{-1}(u_1), \ldots, \mathbf{T}_\nu^{-1}(u_n); \mathbf{R}, \nu\right). \tag{2.2}$$

The Student's t copula is defined by $n(n-1)/2$ correlation parameters and additionally the degree of freedom parameter, that is, $n(n-1)/2+1$ parameters in total. In the case of an exchangeable correlation matrix with $\rho_{jk} = \rho \in (-1/(n-1), 1)$ for all $j, k = 1, \ldots, n$, $j \neq k$, the Student's t copula is exchangeable itself and is specified by only two parameters. Moreover if $\rho_{jk} = 1$ for all $j, k = 1, \ldots, n$, $j \neq k$, we represent comonotonicity, while in the bivariate case, we have countermonotonicity for $\rho_{12} = -1$. For $\nu < \infty$ we cannot represent the independence case.

In the bivariate case, the copula depends only on the off-diagonal parameter of the correlation matrix, i.e., $\rho = \rho_{12}$ and ν. In this case the Kendall's τ can be represented in term of one of these parameters as

$$\tau(\rho) = \frac{2}{\pi} \arcsin(\rho),$$

while Spearman's ρ_S has no closed-form expression.

Like the bivariate Gaussian copula, it is symmetric, but differently from that, it has tail dependence, given by

$$\lambda_L(\rho, \nu) = \lambda_U(\rho, \nu) = 2T_{\nu+1}\left(-\sqrt{\nu+1}\sqrt{\frac{1-\rho}{1+\rho}}\right).$$

We observe that the Student's t copula converges to the Gaussian copula if $\nu \to \infty$. Hence the larger the degree of freedom is, the weaker the tail dependence. Moreover we emphasize that although the Student's t copula is tail-dependent, the tail dependence among pairs of variables is symmetric in both tails as a consequence of the radial symmetry of the elliptical family, and it is the function of only one parameter. An extension of the standard Student's t copula which allows having additional flexibility in the tails is proposed by Luo and Shevchenko [23].

2.2.2 *Archimedean Copulas*

The other major family of copulas that is often used in applications is the so-called class of Archimedean copulas. In contrast to the elliptical copulas, it is not easy to explain how to introduce them directly from multivariate distributions.

Let $\phi(x)$ be a function, and call it the generator of the Archimedean copula function. We will (shortly) digress below on the technical requirements this function must satisfy to be a generator. Now we use it to define the class of Archimedean copulas as follows

$$C(\mathbf{u}) \equiv \phi^{-1}\left(\sum_{i=1}^{n}\phi(u_i)\right). \tag{2.3}$$

To provide a general way to define the requirements which must be met to generate an Archimedean copula function, let us focus on the inverse of the generator $\psi(\cdot) \equiv \phi^{-1}(\cdot)$. In order to allow for every dimension n, this function is required to be infinitely differentiable, with derivatives of ascending orders of alternate sign, and such that $\psi(0) = 1$ and $\lim_{x\to+\infty} \psi(x) = 0$.[3]

Hence given a n-dimensional Archimedean copula that is absolutely continuous, i.e., $(\phi^{-1})^{n-1}$ exists and is absolutely continuous in $(0, \infty)$, it has a density given by

$$c(\mathbf{u}) = (\phi^{-1})^n(\phi(u_1) + \ldots + \phi(u_n))\prod_{i=1}^{n}\phi'(u_i),$$

[3]These features define a famous class of functionals, i.e., the family of *Laplace transforms*.

where $u_i \in [0, 1], i = 1, 2$. See McNeil and Nešlehová [24] for further details.

The Kendall function of a n-dimensional Archimedean copula is very complex and is given in term of the generator and the highest order of derivatives of its inverse (see Barbe et al. 1). In the bivariate case, we have

$$K(z, \phi) = z - \frac{\phi(z)}{\phi'(z)},$$

where $z \in [0, 1]$. Similarly the Kendall's τ can be written as

$$\tau(\phi) = 1 + 4 \int_0^1 \frac{\phi(z)}{\phi'(z)} dz,$$

while there is not a closed-form expression for the Spearman's ρ_S.

The tail dependence coefficients can be represented as follows

$$\lambda_L(\phi) = \lim_{t \to \infty} \frac{\phi^{-1}(2t)}{\phi^{-1}(t)}$$

$$\lambda_U(\phi) = 2 - \lim_{t \to 0} \frac{1 - \phi^{-1}(2t)}{1 - \phi^{-1}(t)}.$$

Finally we observe that Archimedean copulas are exchangeable, meaning that all pairs of variables are equally distributed. This strict assumption is overcome by their non-exchangeable extension, namely, by the *Hierarchical Archimedean copulas* that will be introduced in the following section.

We will now delve into the details of the most common Archimedean copulas which exhibit different properties especially with respect to their tail dependencies.

For example, $\psi(t) \equiv (1 + \theta t)^{-1/\theta}$ that is the Laplace transform of the *Gamma* distribution, yields

$$\phi(t) = \frac{1}{\theta}\left(t^{-\theta} - 1\right), \tag{2.4}$$

and Equation (2.3) produces the most famous among Archimedean copulas, that is, the *Clayton copula*

$$C(\mathbf{u}) = \max\left(\left(\sum_{i=1}^{n} u_i^{-\theta} + n - 1\right)^{-1/\theta}, 0\right). \tag{2.5}$$

The limiting cases of Clayton copula are independence if $\theta \to 0$, comonotonicity if $\theta \to \infty$, and bivariate countermonotonicity if $\theta \to -1$. The corresponding Kendall's τ is given by

$$\tau(\theta) = \frac{\theta}{\theta + 2}.$$

Finally we point out that Clayton copula is a radially asymmetric copula and thus tail asymmetric, showing lower tail dependence for $\theta > 0$. In fact the tail coefficients are given by

$$\lambda_L(\theta) = 2^{-1/\theta}$$
$$\lambda_U(\theta) = 0.$$

Other famous example in the class of Archimedean copulas is represented by the *Frank copula*, obtained from the generator

$$\phi(t) = -\ln\left(\frac{\exp(-\theta t) - 1}{\exp(-t) - 1}\right). \tag{2.6}$$

The *Frank copula* is then defined as

$$C(\mathbf{u}) = -\ln\left(1 + \frac{\prod_{i=1}^{n}(\exp(-\theta u_i) - 1)}{\exp(-\theta) - 1}\right). \tag{2.7}$$

The Frank copula has the same properties as the Gaussian copula, and it is indeed very close to it in shape. It converges to independence for $\theta \to 0$ and to comonotonicity for $\theta \to \infty$, while in the bivariate case if $\theta \to -\infty$ covers the countermonotonicity.
The Kendall's τ and the Spearman's ρ_S can be written as

$$\tau(\theta) = 1 + \frac{4}{\theta}(D_1(\theta) - 1)$$
$$\rho_S(\theta) = 1 - \frac{12}{\theta}(D_1(\theta) - D_2(\theta)),$$

where $D_k(x)$ stands for the Debye function, i.e.

$$D_k(x) = \frac{k}{x^k}\int_0^x \frac{t^k}{e^t - 1}\,dt\ ,$$

where $x \in \mathbb{R} \setminus \{0\}$, $k \in \mathbb{N}$.

In the bivariate case, the Frank copula is radially symmetric while it looses this property for $n > 2$, and it is not tail-dependent, i.e., $\lambda_L(\theta) = \lambda_U(\theta) = 0$.

Another famous Archimedean copula is generated by $\phi(t) \equiv (-\ln t)^\alpha$ which yields the *Gumbel family*:

$$C(\mathbf{u}) = \exp\left(-\left(\sum_{i=1}^{n}(-\ln u_i)^\theta\right)^{1/\theta}\right), \tag{2.8}$$

whose limiting cases are independence for $\theta = 1$ and comonotonicity if $\theta \to \infty$. In contrast to the families of copulas reviewed so far, which at least in the bivariate case encompasses the cases of independence, perfect positive and negative dependence, the Gumbel copula is limited to positive dependence only. In the jargon of copula functions, we say it is not *comprehensive*.

The Kendall's τ is given by

$$\tau(\theta) = 1 - \frac{1}{\theta},$$

while there is no closed-form expression of Spearman's ρ_S.

Gumbel copula has upper tail dependence. Finally we point out that Gumbel copula is a radially asymmetric copula and thus tail asymmetric, showing upper tail dependence given by

$$\lambda_L(\theta) = 0$$
$$\lambda_U(\theta) = 2 - 2^{1/\theta}.$$

Figure 2.6 reports the scatter plots of the Clayton, Frank, and Gumbel copula highlighting the tail dependence property of these families.

Another Archimedean copula which shows the same tail dependence coefficients as the Gumbel copula is the *Joe family* whose generator is $\phi(t) \equiv -\log(1-(1-t)^{\theta})$. The corresponding copula is then

$$C(\mathbf{u}) = 1 - \left(1 - \prod_{i=1}^{n}(1-(1-u_i)^{\theta})\right)^{1/\theta},$$

where $u_i \in [0, 1]$, $i = 1, \ldots, n$ and $\theta > 1$. The limiting cases are the independence for $\theta \to 1$ and the comonotonicity for $\theta \to \infty$.

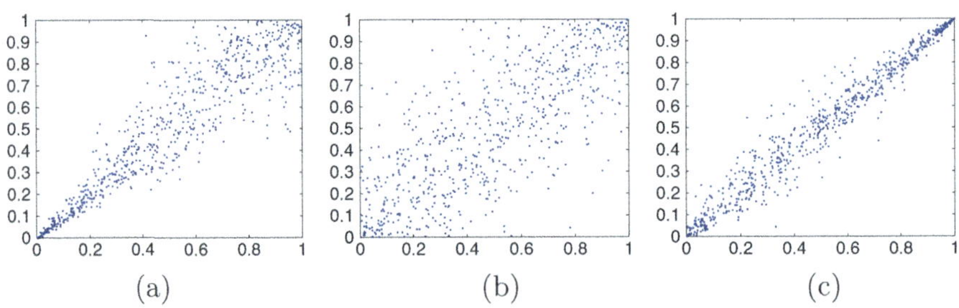

Fig. 2.6 Scatter plots of 500 data simulated from the bivariate Clayton, Frank and Gumbel copula with $\theta = 5$. (**a**) Clayton copula. (**b**) Frank copula. (**c**) Gumbel copula

We recall that there exists a closed-form representation of his Kendall's τ in terms of the digamma function, which corresponds to the logarithmic derivative of the gamma function.

2.2.3 Extreme-Value Copulas

Extreme-value copulas, introduced by Pickands [28], can be defined as the asymptotic limit of the component-wise maxima. Given d-independent copies of a n-dimensional random vector $\mathbf{X}$ with copula C_0, we define $M_{d,i} = \max\{X_{1i}, \ldots, X_{di}\}$ that is the maximum of the ith component over n copies. Then the copula of $M_n = (M_{d,1}, \ldots, M_{d,n})$ is

$$C_{0(n)}(\mathbf{u}) = C_0(\mathbf{u}^{1/n})^n,$$

for all $u_i \in [0,1], i = 1, \ldots, n$. C is an extreme-value copula if there exists a copula C_0 which lies in the domain of attraction of C, i.e.,

$$C_{0(n)}(\mathbf{u}) \xrightarrow[n\to+\infty]{} C(\mathbf{u}), \forall \mathbf{u} \in [0,1]^n.$$

In the bivariate case, an extreme-value copula is uniquely defined by the Pickands function $A : [0,1] \to [0.5,1]$ which is convex and such that $\max\{t, 1-t\} \leq A(t) \leq 1, \forall t \in [0,1]$. Hence the corresponding extreme-copula can be written as

$$C(u_1, u_2, A) = \exp\left(\log(u_1 u_2) A\left(\frac{\log u_2}{\log(u_1 u_2)}\right)\right), \quad (u_1, u_2) \in [0,1]^2.$$

On the other hand , if C is a bivariate extreme-value copula, there exists a Pickands function A which allows the above representation of the copula. The limiting cases are represented by independence if $A(t) = 1, \forall t$, and comonotonicity if $A(t) = \max\{t, 1-t\}$.

Finally we observe that an extreme-value copula is symmetric if and only if the Pickands function is symmetric about 0.5, i.e., $A(t) = A(1-t), \forall t \in [0,1]$. Moreover the bivariate extreme-value copula has an explicit Kendall function which depends on the Pickands function, as follows:

$$K(z, A) = z(1 + (\tau(A) - 1)\log z), \quad z \in [0,1],$$

where the Kendall's $\tau(A)$ of the extreme-value copula is defined as

$$\tau(A) = \int_0^1 \frac{t(1-t)}{A(t)} dA'(t),$$

assuming that the right-hand derivative of A on $[0, 1)$, noted as A', exists. Similarly the Spearman's $\rho_S(A)$ can be represented in term of the Pickands function:

$$\rho_S(A) = 12 \int_0^1 \frac{1}{(1 + A(t))^2}\, dt - 3.$$

The extreme-value copulas are in general lower tail-independent except in the comonotonicity case, where $A(0.5) = 0.5$, and they are upper tail-dependent, but they are not radially symmetric except in the limiting cases. In fact the tail dependence coefficients can be written as

$$\lambda_L(A) = \begin{cases} 1 \text{ if } A(0.5) = 0.5 \\ 0 \quad \text{otherwise} \end{cases}$$

$$\lambda_U(A) = 2(1 - A(0.5)).$$

It is proved that the Gumbel copula is the only copula that is both Archimedean and an extreme-value copula characterized by the following Pickands function which is symmetric about 0.5:

$$A(t) = (t^\theta + (1 - t)^\theta)^{1/\theta}, \quad t \in [0, 1].$$

Archimedean copulas lie in the domain of attraction of Gumbel copula if there exists

$$\lim_{s \to 0} \frac{s\phi'(1 - s)}{\phi(1 - s)} \in [1, \infty]$$

and then produce radially symmetric copulas.

An extension of Gumbel copula which allows for asymmetry is the *Tawn copula* whose Pickands function is given by

$$A(t, \theta, \psi_1, \psi_2) = (1-\psi_2)(1-t)+(1-\psi_1)t+((\psi_2 t)^\theta+(\psi_1(1-t))^\theta)^{1/\theta}, \quad t \in [0, 1]$$

where $\theta \geq 1$ and $\psi_1, \psi_2 \in [0, 1]$. We observe that if $\psi_1 = \psi_2 = 1$, the Tawn copula reduces to the Gumbel copula, while if $\psi_1 \neq \psi_2$ it shows an asymmetric behavior. Moreover the independence case corresponds to the assumption $\theta = 1$ or $\psi_1 = 0$ or $\psi_2 = 0$,while for $\theta \to \infty$ it converges to the *Marshall-Olkin copula*.

2.3 Pure Hierarchical Copulas

Hierarchical Archimedean copulas have been introduced in Savu and Trede [29] to offset the major flaw of this class coming from exchangeability, i.e., the assumption that the dependence parameter is the same across several pairs of assets. In fact, it is easy to verify that Archimedean copulas lead to the following result

$$\begin{aligned} C(1,\ldots,1,u_i,1,\ldots,1,u_j,1,\ldots,1) &= \phi^{-1}(\phi(u_i)+\phi(u_j)+\phi(1)+\ldots+\phi(1)) \\ &= \phi^{-1}(\phi(u_i)+\phi(u_j)), \forall i,j \end{aligned}$$

where ϕ is the generator of the Archimedean copula. A way to generalize this representation, allowing for $C(u_i,u_j) \neq C(u_m,u_n), i \neq m, j \neq n$, is to impose a *hierarchical* representation. Hierarchical Archimedean copulas can be fully or partially nested depending on the kind of aggregation we use. For example, a *fully nested* copula in dimension $n = 3$ corresponds to

$$C(\mathbf{u}) = \phi_2^{-1}\left(\phi_2(u_3) + \phi_2 \circ \phi_1^{-1}(\phi_1(u_2) + \phi_1(u_1))\right)$$

where ϕ_1 represents the generator explaining the dependence between the first two variables; ϕ_2 stands for the highest level of dependence, i.e., between the previous couple and the third variable and $\circ$ is the composition operator. Note that in this three-dimensional case, we have two nesting levels, meaning that a n-dimensional fully nested Archimedean copula is able to capture $n-1$ different pairwise dependences. Otherwise, C is called *partially nested* Archimedean copula (see Hofert [18], Hering et al. [17], Okhrin et al. [26] and Okhrin et al. [27], Choros et al. [7], as suggested references about the hierarchical Archimedean family).

Although fully and partially nested Archimedean copulas are summarized as nested or hierarchical Archimedean copulas, we classify them, distinguishing between

1. *Partially exchangeable copulas*. It is based on grouping aggregation that represents the second level of the hierarchy and a final aggregation between groups, corresponding to the first level of the hierarchy. It is a particular partially nested copula where the exchangeability between groups, i.e., the partial exchangeability property, is preserved. In n-dimension and with pair aggregation, the resulting copula is

$$\begin{aligned} &PE-HC(\mathbf{u}) \\ &\equiv \phi_s^{-1}\left(\phi_s \circ \phi_1^{-1}(\phi_1(u_1)+\phi_1(u_2)) + \ldots + \phi_s \circ \phi_n^{-1}(\phi_n(u_{n-1})+\phi_n(u_n))\right), \end{aligned}$$

where ϕ_s represents the generator of the Archimedean copula explaining the dependence between couples, $\phi_j, j = 1,\ldots,n$ represent the dependence within couples and $\circ$ is the composition operator. This representation uses different generators $\phi_1,\ldots,\phi_n$ for the couples (u_1,u_2), (u_2,u_3) , ..., (u_{n-1},u_n) and a generator $\phi_s \circ \phi_i^{-1}, i = 1,\ldots,n$ to model the dependence of the pairs (u_1,u_3), $(u_2,u_4), \ldots (u_{n-2},u_n)$. For the generator $\phi_s \circ \phi_i^{-1}, i = 1,\ldots,n$ to be well-defined, the dependence parameter must be decreasing from the bottom to the highest hierarchical level as proved in Embrechts et al. [13]. So, denoting by

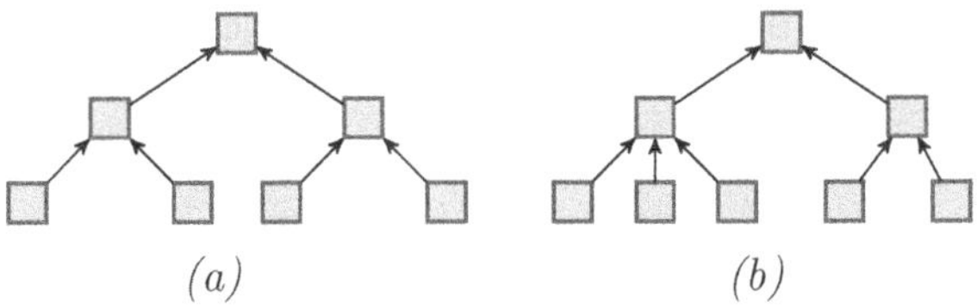

Fig. 2.7 Partially exchangeable hierarchical aggregations

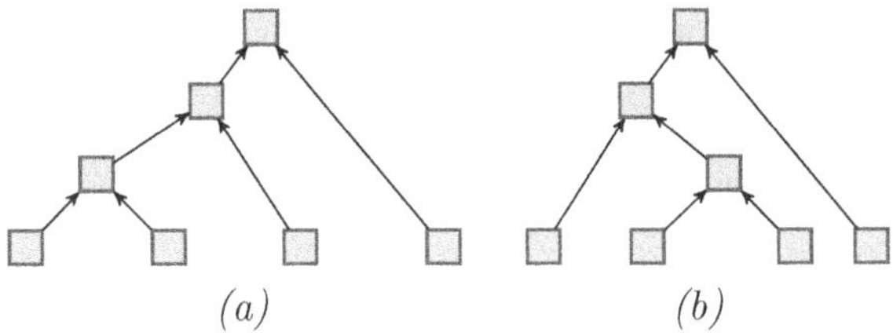

Fig. 2.8 Not exchangeable hierarchical aggregations

α_i the dependence parameter of the generator ϕ_i, in our case it must be $\alpha_s \leq \max\{\alpha_i, i = 1, \ldots, n\}$. Moreover it is possible to choose generators of different Archimedean families. Note however that some families are incompatible with each other and will never form a valid copula. As, for example, Gumbel and Clayton generators are incompatible in hierarchical models (see Hofert 18).

This kind of aggregation is represented in Fig. 2.7 assuming a four-dimensional setting. In Fig. 2.7a, the aggregations have been represented as couples, but they could be generalized assuming a second level of aggregation within groups whose cardinality is not necessarily equal to 2 as we can see in Fig. 2.7b.

2. *Not exchangeable copulas*. This kind of hierarchy is recovered through a nested aggregation that proceeds by a first group aggregation and a series of single aggregations as represented in Fig. 2.8a assuming a four-dimensional setting. In this example we have a fully nested copula, but we can imagine also a partially nested case that proceeds by a nested aggregation by groups while by single ones (see Fig. 2.8b).

In n-dimensions the resulting copula is

$$NE - HC(\mathbf{u}) \equiv \phi_{n-1}^{-1}\left(\phi_{n-1}(u_n) + \phi_{n-1} \circ \phi_{n-2}^{-1}\left(\phi_{n-2}(u_{n-1})\right) + \ldots + \right.$$
$$\left. + \phi_{n-2} \circ \phi_{n-3}^{-1}\left(\phi_{n-3}(u_{n-2}) + \phi_2 \circ \phi_1^{-1}(\phi_1(u_2) + \phi_1(u_1))\right)\right).$$

For the nested copula to be well-defined, we must assume the dependence parameter is decreasing from ϕ_{n-1} to ϕ_1, i.e., the highest level of the hierarchy must have the feeblest dependence. Moreover also for the nested aggregation generating not exchangeable copulas, it would be possible to choose generators coming from different Archimedean families, if they are compatible according to the meaning specified by Hofert [18].

Other kind of hierarchical and not exchangeable dependence is represented by the class of *Vine copulas* which are multivariate copulas defined by a number of bivariate copulas. In this case a multivariate distribution is defined by recursive conditioning of pair copulas. Bedford and Cooke [2] introduced vines as a graph theoretical model to organize different pair copulas aggregation. More details on vine copulas can be found in Czado [10] and Kurowicka and Joe [22].

2.4 Hierarchical Grouping Copulas

In this section we generalize the notion of *pure hierarchical copula* assuming to aggregate groups instead of single variable. In this case we usually talk about *hierarchy* among classes. The issue of non-exchangeability becomes unavoidable if one addresses the dependence problem in higher dimensions. In this respect the first intuitive idea that would come to mind is to arrange the variables in clusters and to impose a dependence structure between the clusters. This idea leads to the concept of hierarchical grouping copulas.

Moreover we stress that in this kind of aggregation, the way the classes are defined becomes relevant in order to recover the multivariate distribution. We can imagine to define several classes based on specified characteristics of the variables and through one of the clustering method discussed in Chap. 1. In this kind of approach which leads to the so-called clusterized copulas, we need to aggregate groups of variables which can be assumed to be similar each other based on the similarity features used in the clustering phase. We disentangle two kinds of clusterized copulas, i.e., the *homogeneous* class where the variables inside the same group are assumed to be homogeneous and the *not homogeneous* or *hierarchical* where we have a hierarchy also inside the groups. A special case of the homogeneous class where we assume to have only one level of dependence leads to an aggregation which is not hierarchical but which may be useful in order to reduce the complexity of an high-dimensional problem.

2.4.1 Clusterized Homogeneous and Clusterized Hierarchical Copulas

The class of *clusterized homogeneous copulas* (CHC for short in the following) had been introduced in Bernardi and Romagnoli [5]. They are grouping copulas whose classes, recovered through a clustering method, are assumed to be homogeneous and are related by a dependency which is possibly hierarchical. Therefore every class is characterized by a perfect positive dependence on his centroid, while the aggregation *between* classes may be represented by more than one level of dependence.

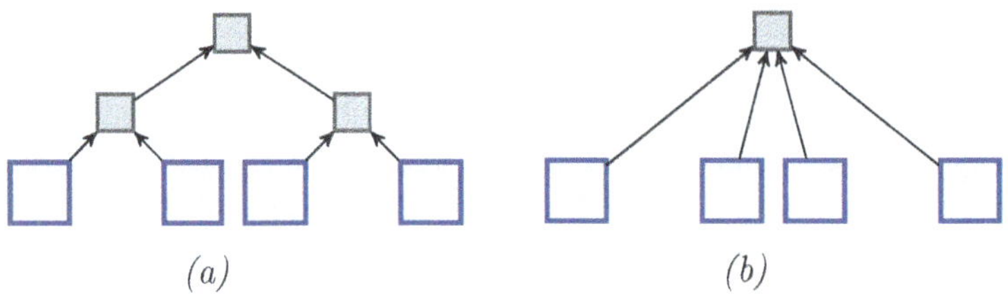

Fig. 2.9 Clusterized Homogeneous copulas: (**a**) 2-CHC and (**b**) 1-CHC

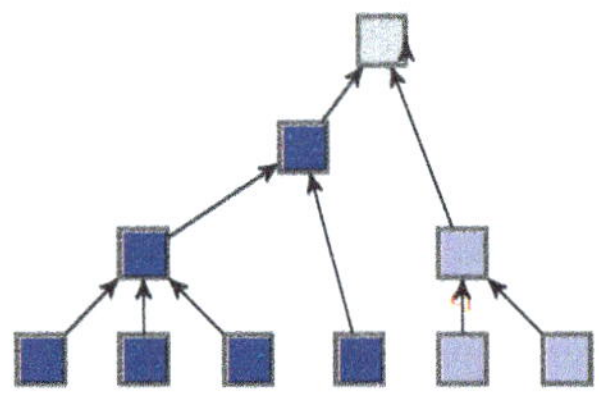

Fig. 2.10 Clusterized hierarchical copulas: three levels of dependence

An important special copula of this class is the one-level CHC copula.

Definition 2.4.1 (1-CHC) A one-level-clusterized homogeneous copulas is a grouping copula based on a clustering method, having only one level of dependence and whose groups are assumed to be homogeneous.

We point out that if the dependence level is only one, this kind of copula cannot represent a pure hierarchy, but it is useful to diminish the dimension of the problem. We remark that a no hierarchical and no clusterized copula can be seen as a 1-CHC with single groups.

In Fig. 2.9 is represented a CHC with two levels of dependence (Fig. 2.9a) and with only one level of dependence (Fig. 2.9b); the shadow blue squares stand for the homogeneous groups which are represented by the corresponding centroids. As we can observe, the cardinality of the homogeneous groups is reduced to a single unit, i.e., the centroid, allowing to manage original high-dimensional problems. In Fig. 2.10 is plotted a clusterized hierarchical copula; the dark blue and the pale blue squares represent any single component of two groups whose cardinality is four and two, respectively, with totally three levels of dependence, two and one *within*, respectively, and one *between*. As we can observe in this case the clustering doesn't allow to reduce the complexity of the problem because the grouping aggregates the variables without reducing their cardinality.

In general to implement a clusterized homogeneous copula, it is necessary to represent the dependence structure of the given set of variables by a groups' aggregation. Then once a clustering method is selected and a copula function is chosen, we can estimate from data the dependence parameter α of that copula through a parametric approach. Therefore to reduce the complexity of the problem, we can represent the dependence structure of the set of variables with a CHC constructed through a homogenization procedure of the clusters.

Our aim is to consider the set of data and assume to partition them into so-called homogeneous clusters such that patterns within a cluster are more similar to each other than patterns belonging to different clusters. We refer to Chap. 1 for a discussion about the choice of a clustering method that best fits our requests. The clustering method assembles in groups, which must be considered homogenous to simplify the computation in case of high-dimensional problems. To this aim we develop a procedure, along the lines of the Diversity Score approach suggested in Cifuentes and O'Connor [8], Cifuentes et al. [9] and applied in Schorin and Weinrich [30] and in Duffie and Gârleanu [12], allowing us to consider that the variables can be considered homogenous. In order to do it we must compensate the homogenous assumption with a change in the dependence parameter inside the group or in the size of the group. In order to compensate the homogenous assumption with a change in the number of variables inside the group, we assume the groups' cardinality of the clusterized copula function or the copula's parameter[4] to be *equivalent* in the meaning adopted in Bernardi and Romagnoli [3].

If we compensate with the dependence parameter, we have the following:

Definition 2.4.2 The Equivalent Dependence Parameter (EQDP) for the copula function C of group $i, i = 1, \ldots, n$ is α_i such that

$$C(\mathbf{u}^i, \widehat{\alpha_i}) = C(\mathbf{c}_i, \alpha_i),$$

where $u^i_j, j = 1, \ldots, m_i$ are the heterogenous marginals of the group i of dimension m_i, $C(\mathbf{c}_i, \alpha_i)$ is an m_i-variate copula with homogeneous marginals c_i equal to the centroid of the i-th cluster, and $\widehat{\alpha_i}$ is the standard estimation of the dependence parameter within the heterogenous group for the same copula function.

On the other hand, if we decide to change the number of the variables to be considered into the homogenous group, we have the following:

Definition 2.4.3 The Equivalent Number of Variables (ENV) for the copula function C of group $i, i = 1, \ldots, n$ is the dimension $\hat{m}_i$ of copula C after the homogeneous approximation:

$$C^{(m_i)}(\mathbf{u}^i, \widehat{\alpha_i}) = C^{(\hat{m}_i)}(\mathbf{c}_i, \widehat{\alpha_i}),$$

where $u^i_j, j = 1, \ldots, m_i$ are the heterogenous marginals of group i, $C^{(\hat{m}_i)}(\mathbf{c}_i, \widehat{\alpha_i})$ is an $\hat{m}_i$-variate copula with homogeneous marginals c_i equal to the centroid of the i-th cluster, and $\widehat{\alpha_i}$ is the standard estimation of the dependence parameter within the heterogenous group for the same copula function.

[4]We talk about copula's parameter because in this procedure we work with hierarchical copulas and so with Archimedean ones that are characterized by only one parameter.

On the other hand, if the grouping doesn't aim to work with homogeneous classes, defined by a clustering method adjusted in order to compensate for the homogeneous assumption, we lead to a hierarchy also inside every class. In this case the clustering doesn't allow to diminish the complexity of the problem as we did by the homogeneous transform discussed before. This kind of aggregation defines the class of *clusterized hierarchical copulas* (CHY for short).

Finally we can detail the class of clusterized copulas considering their limit in case of large classes, i.e., classes whose cardinality is greater to a given threshold.[5] The limit of the CHC is a new class of copula called *Hierarchical Limiting copula*, (HLC for short); it was introduced in Bernardi and Romagnoli [4], and it can be considered a limiting version of a CHC where we assume to have a proper hierarchy, i.e., more than one level of dependence. The idea is here to clusterize the variables and then to apply a multiple level CHC whose marginal cdfs are recovered with an approximated model useful for large cardinalities. These approximated models that will be discussed in Part II of the book in the contest credit risk applications are called in literature *Pure limiting models* (PL model for short).

In the same way, the limit of the class of *clusterized hierarchical copulas* in case we have a subset of large classes, allows us to introduce a new family of copula called *Hierarchical hybrid copula* (HYC for short). This class is characterized by large classes, whose marginal cdfs are recovered through an approximated model for large cardinalities and some small classes which are hierarchical; this copula is said to be hybrid because it unifies two different kind of aggregation *within*, i.e., the pure hierarchical from one side to the limiting models for the other side.

2.4.2 Hierarchical Kendall Copulas

The class of Hierarchical Kendall copulas have been introduced in dos Anjos and Kolev [11] and refer to a special kind of clusterized copulas. The name *Hierarchical Kendall copula* stresses the central role the Kendall distribution function plays in the model formulation.

Definition 2.4.4 (Hierarchical Kendall copula) Let $U_1, \ldots, U_n \sim U(0,1)$ and let $C_1, \ldots, C_n$ be copulas of dimension $n_1, \ldots, n_d$, respectively, $n_i \geq 1, i = 1, \ldots, d$, and $n = \sum_{i=1}^{d} n_i$. Moreover let $K_1, \ldots, K_d$ denote the Kendall distribution functions corresponding to $C_1, \ldots, C_d$, respectively. We define $m_i = \sum_{j=1}^{i} n_j, i = 1, \ldots, d$, and $m_0 = 0$ as well as $\mathbf{U}_i = (U_{m_{i-1}+1}, \ldots, U_{m_i})'$ and $V_i = K_i(C_i(\mathbf{U}_i)), i = 1, \ldots, d$. Under the assumptions that

1. $\mathbf{U}_1, \ldots, \mathbf{U}_d$ are mutually independent conditionally on $(V_1, \ldots, V_d)'$.
2. The conditional distribution of $\mathbf{U}_i|(V_1, \ldots, V_d)'$ is the same as the conditional distribution of $\mathbf{U}_i|V_i, \forall i = 1, \ldots, d$, that is, $F_{\mathbf{U}_i|V_1,\ldots,V_d} = F_{\mathbf{U}_i|V_i}, \forall i \in \{1, \ldots, d\}$.

[5] This threshold is fixed coherently with the problem at hand.

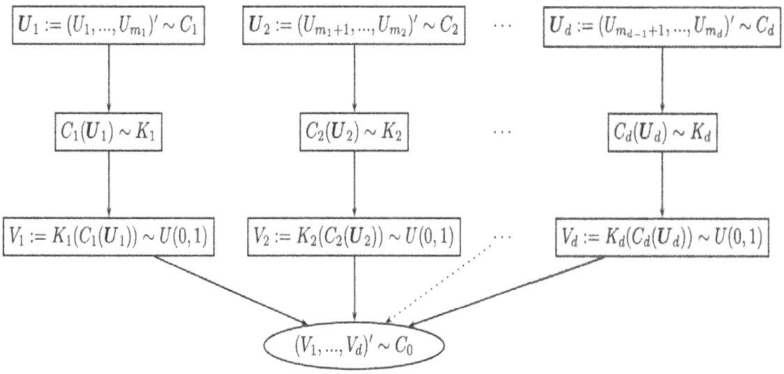

Fig. 2.11 Graphical representation of a Hierarchical Kendall copula

The random vector $(U_1, \ldots, U_n)'$ is said to be distributed according to the hierarchical Kendall copula $C_{\mathcal{K}}$ with nesting copula C_0 and cluster copulas $C_1, \ldots, C_d$ if

1. $\mathbf{U}_i \sim C_i, \forall i \in \{1, \ldots, d\}$
2. $(V_1, \ldots, V_d)' \sim C_0$.

The intuition behind the two previous assumptions is that, given the information of the nesting variables $(V_1, \ldots, V_d)$, the clusters $(\mathbf{U}_1, \ldots, \mathbf{U}_d)$ are independent of each other and also of other nesting variables, since the dependence among the clusters is explained through the variables $(V_1, \ldots, V_d)$. In other words, the variables $(V_1, \ldots, V_d)$ can be interpreted as unobserved factors, whose joint behavior determines the dependence of the different clusters. In finance, such factors may be, for instance, industry sectors. In Fig. 2.11 we have a useful representation of the Kendal copulas aggregation. We observe that C_0 is in general not the copula of $(U_1, \ldots, U_n)'$ but of $(V_1, \ldots, V_d)'$, which are uniform random variables due to $C_i(\mathbf{U}_i) \sim K_i, \forall i = 1, \ldots, d$. On the other hand, the nesting copula C_0 and the cluster copulas $C_1, \ldots, C_d$ can be chosen independently. They can be arbitrary copulas such as common Archimedean or elliptical copulas or any other class of copulas.

2.5 Volume of an n-Dimensional Copula

In this section we deal with the concept of volume of a copula function which is diffusely used in the aggregation algorithm depicted in the second part of this book.

In a bivariate setting, the volume of a copula function C between two points $\mathbf{u} = [u_1, u_2]$ and $\mathbf{v} = [v_1, v_2]$, where $u_i \leq v_i, i = 1, 2$, is defined as

$$V_C(\mathbf{S}) = C(v_1, v_2) - C(u_1, v_2) - C(v_1, u_2) + C(u_1, u_2),$$

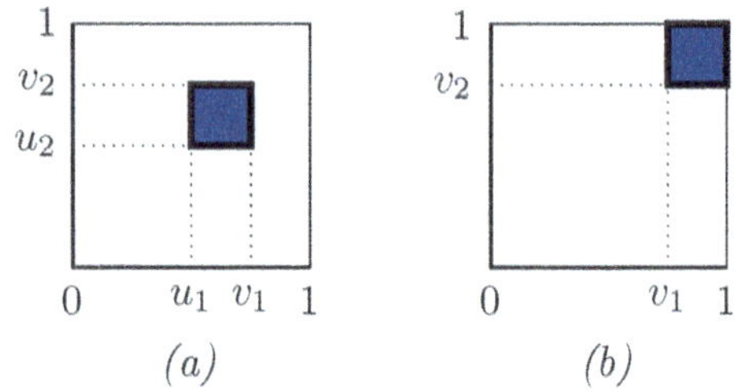

Fig. 2.12 Graphical representation of the volume of copula C in the boxes $\mathbf{S} = [\mathbf{u}, \mathbf{v}]$, and $\mathbf{S} = [\mathbf{v}, \mathbf{1}]$

where $\mathbf{S} = [\mathbf{u}, \mathbf{v}]$. This representation is easily understood if you look at Fig. 2.12a where the volume is geometrically represented.

An interesting interpretation of the volume concerns a particular box defined as $\mathbf{S} = [\mathbf{v}, \mathbf{1}]$ where $\mathbf{1}$ is a two-dimensional unit vector. Through the same representation used before, we have

$$\begin{aligned} V_C(\mathbf{S}) &= C(1,1) - C(1,v_2) - C(v_1,1) + C(v_1,v_2) \\ &= 1 - v_2 - v_1 + C(v_1,v_2), \end{aligned}$$

that is geometrically represented in Fig. 2.12b.

This volume, which represents a particular copula $\bar{C}$, called *survival copula* function, has marginals equal to the complement of marginals of copula C, i.e.,

$$\bar{C}(1-v_1, 1-v_2) = V_C(\mathbf{S}),$$

where $\mathbf{S} = [\mathbf{v}, \mathbf{1}]$.

Computing the volume in higher dimensions can be much more complex. In cases where analytical computation of the copula is feasible, an interesting question is whether the volume can also be computed exactly and whether such analytical computation is feasible. In Cherubini and Romagnoli [6], a simple algorithm (CR in the following) to address this problem is provided. To understand this algorithm it is fundamental to give the following representation of copula's volume.

Proposition 2.5.1 **(C-volume)** *Given a n-dimensional copula function $C(\mathbf{u})$ with $\mathbf{u} \in [0,1]^n$, the C-volume of the box $\mathbf{S} = [\mathbf{u}, \mathbf{v}]$ with $\mathbf{v} \in [0,1]^n$ such that $\mathbf{u} \leq \mathbf{v}$ can be represented as*

$$V_C(\mathbf{S}) = \sum_{i=0}^{2^n-1} (-1)^{k(i)} C(\mathbf{c}_n(\mathbf{p}_n(i))),$$

where $\mathbf{p}_n(i)$ denotes the n-dimensional binary representation of i, $k(i) = |\mathbf{p}_n(i)|$ counts the number of elements equal to one in $\mathbf{p}_n(i)$, and $\mathbf{c}_n(\mathbf{p}_n(i))$ is a n-

dimensional vector such that $c_j = v_j$ *if* $p_j = 0$ *and* $c_j = u_j$ *if* $p_j = 1$*, where* u_j, v_j, p_j*, and* c_j *denote the* j*th element of the corresponding vector.*

Proof For $n = 2$ it is immediate to prove the statement

$$\begin{aligned} V_C(\mathbf{S}) &= \sum_{i=0}^{3}(-1)^{k(i)}C(\mathbf{c}_2(\mathbf{p}_2(i))) \\ &= C(\mathbf{c}_2(0,0)) - C(\mathbf{c}_2(0,1)) - C(\mathbf{c}_2(1,0)) + C(\mathbf{c}_2(1,1)) \\ &= C(v_1, v_2) - C(v_1, u_2) - C(u_1, v_2) + C(u_1, u_2). \end{aligned}$$

We prove it by induction assuming that the statement is verified for n and proving it for $n + 1$. We observe that $\mathbf{p}_{n+1}(i)$ can be decomposed as

$$\mathbf{p}_{n+1}(i) = (0, \mathbf{p}_n(i)) \vee (1, \mathbf{p}_n(i))$$

and then it follows that

$$\begin{aligned} V_C(\mathbf{S}) &= \sum_{i=0}^{2^n-1}(-1)^{k(i)}C(\mathbf{c}_{n+1}(0, (\mathbf{p}_n(i)))) + \sum_{i=0}^{2^n-1}(-1)^{k(i)}C(\mathbf{c}_{n+1}(1, (\mathbf{p}_n(i)))) \\ &= \sum_{i=0}^{2^{n+1}-1}(-1)^{k(i)}C(\mathbf{c}_{n+1}(\mathbf{p}_n(i))). \end{aligned}$$

□

The previous representation can be specialized for *exchangeable copulas*. These are copulas where the position of the marginals can be exchanged without the aggregated result changes, i.e., $C(u, v) = C(v, u), \forall u, v$ in a 2 dimensional setting.

Proposition 2.5.2 (exchangeable C-volume) *Given a* n*-dimensional exchangeable copula function* $C(\mathbf{u})$ *with* $\mathbf{u} \in [0, 1]^n$*, the* C*-volume of the box* $\mathbf{S} = [\mathbf{u}, \mathbf{v}]$ *with* $\mathbf{v} \in [0, 1]^n$ *such that* $\mathbf{u} \le \mathbf{v}$ *can be represented as*

$$V_C(\mathbf{S}) = \sum_{i=0}^{n}(-1)^i \binom{n}{i} C(\mathbf{c}_n(\bar{\mathbf{p}}_n(i))),$$

where $\bar{\mathbf{p}}_n(i)$ *denotes the* n*-dimensional vector with the* i*th elements equal to* 1 *and the other* $n - i$ *elements equal to* 0.

Proof For the exchangeability of the copula function the binary representation of number i is equivalent to the binary representations of others numbers with the same number of ones, that is $\binom{n}{i}$. Then from Proposition 2.5.1 we have

$$V_C(\mathbf{S}) = C(\mathbf{c}_n(\mathbf{p}_n(0))) - \binom{n}{1} C(\mathbf{c}_n(\mathbf{p}_n(1))) + \binom{n}{2} C(\mathbf{c}_n(\mathbf{p}_n(2)))$$

$$+ \binom{n}{3} C(\mathbf{c}_n(\mathbf{p}_n(3))) + \ldots (-1)^{k(2^n-1)} C(\mathbf{c}_n(\mathbf{p}_n(2^n - 1))),$$

where we sum over the number of ones in the binary representations. □

Computing the C-volume may get very involved for higher dimensions. Cherubini and Romagnoli [6] propose an algorithm that will be referred as CR algorithm in the following, to solve the problem in n-dimensions, with $n \leq 20$. We report it here below.

Remark 2.5.1 (Computation of the survival copula functions) Assume we want to compute the survival version of copula $C(w_1, w_2, \ldots, w_n)$. We use a vector notation denoting $\mathbf{w}$ the vector containing the coordinates of the copula: $\mathbf{w} = \{w_1, w_2, \ldots, w_n\}$. We also use a vector $\mathbf{c}$ that we update at each recursion.

Algorithm (pseudo code)

1. Initialize the vector $\mathbf{c} = [c_1, c_2, \ldots, c_n] = \mathbf{e}$, the unit vector in n-dimension.
2. Set $\hat{C} = 0$.
3. For i = 0 to $2^n - 1$
   . Change i into a binary number, and arrange the elements of such number in vector $\mathbf{p} = [p_1, p_2, \ldots, p_n]$

   $k = 0$

   For j =1 to $f(\log_2(i+1))$

   if $p_j = 1$ then

   $c_j = w_j$

   $k = k + 1$

   end if

   next j

   $\hat{C} = \hat{C} + (-1)^k \, C\,(\mathbf{c})$

   next i
4. Return $\hat{C}$

where $f(\log_2(i+1))$ denotes the function reporting the upper integer value of $\log_2(i+1)$.

Example 2.5.1 To compute the survival copula of $C(\mathbf{u})$ with $\mathbf{u} \in [0, 1]^3$, we consider the C-volume in the box $\mathbf{S} = [\mathbf{u}, 1]$. The algorithm consider the binary representation of all $i = 0, \ldots, 7$, representing vectors $\mathbf{p}_3(i), \forall i$, i.e.,

$$\mathbf{p}_3 = \begin{bmatrix} 0\,0\,0 \\ 1\,0\,0 \\ 0\,1\,0 \\ 0\,0\,1 \\ 1\,1\,0 \\ 1\,0\,1 \\ 0\,1\,1 \\ 1\,0\,1 \\ 1\,1\,1 \end{bmatrix}$$

From $\mathbf{p}_3$ we recover then matrix $\mathbf{c}_3$, substituting the coordinates of the box according to the rule explained before:

$$\mathbf{c}_3 = \begin{bmatrix} 1 & 1 & 1 \\ u_1 & 1 & 1 \\ 1 & u_2 & 1 \\ 1 & 1 & u_3 \\ u_1 & u_2 & 1 \\ u_1 & 1 & u_3 \\ 1 & u_2 & u_3 \\ u_1 & 1 & u_3 \\ u_1 & u_2 & u_3 \end{bmatrix}$$

Finally we compose copula functions evaluated for each rows of matrix $\mathbf{c}_3$ with a sign depending on the integer to which it is a transformed binary representation, i.e.,

$$\begin{aligned} V_C(\mathbf{S}) = {} & C(1, 1, 1) - (C(u_1, 1, 1) + C(1, u_2, 1) + C(1, 1, u_3)) + \\ & + (C(u_1, u_2, 1) + C(u_1, 1, u_3) + C(1, u_2, u_3)) - C(u_1, u_2, u_3) \end{aligned}$$

As for the scalability of the algorithm above, the authors find that good performances in terms of computing time can be achieved even in MatLab for dimensions up to 20. Beyond this dimension, the problem becomes intractable because of physical limits of memory. For problems that exceed this dimension, we need to use different algorithm linked to a particular choice of copula function as we can see in the following.

2.5.1 Clusterized Hierarchical Copulas: CHY-Volume

Here we consider a clusterized copula whose groups are not homogeneous and aggregated taking into account a hierarchy. Bernardi and Romagnoli [3] propose the following representation of the volume of a clusterized hierarchical copula. This algorithm is based on ordered combinatoric distributions of integers instead of the binary representation of i for $i = 1, .., n$. Here CHY is a clusterized hierarchical copula that could be aggregated or nested grouped.

A similar algorithm could be proposed for CHC, where the groups are assumed to be homogeneous. The specialized algorithm will be discussed in the Part II of the book.

Proposition 2.5.3 (clusterized hierarchical copula volume) *Given a clusterized hierarchical copula function CHY with k groups of dimension $m_s, s = 1, .., k$, respectively, such that $\sum_{s=1}^{k} m_s = n$, the volume of the n-dimensional box $\mathbf{S} = [\mathbf{u}, \mathbf{v}]$ with $\mathbf{u}, \mathbf{v} \in [0, 1]^n$, $\mathbf{u} \leq \mathbf{v}$, may be represented as*

$$V_{CHY}(\mathbf{S}) = \sum_{i=0}^{n}(-1)^i \sum_{j=1}^{D^c(i,k)} CHY(\mathbf{c}(\mathbf{p}_{i,k}(j)))$$

where the coordinates defining the box $\mathbf{S}$ are the column aggregation of matrices $\tilde{\mathbf{u}}, \tilde{\mathbf{v}} \in [0, 1]^k \times [0, 1]^{\max\{m_s, s=1,...,k\}}$ whose (s, w)-th element corresponds to the w-th position in the s-th group,[6] *where $D^c(i, k)$ denotes the number of the ordered compatible combinatorial distributions (o.c.c.ds in the following)*[7] *of the integer i into k groups, $\mathbf{p}_{i,k}(j)$ is a $k \times \max(m_s, s = 1, \dots, k)$ matrix of zeros and ones corresponding to the j-th o.c.c.d. of i ones into k groups of dimensions $m_s, s = 1, \dots, k$, $\mathbf{c}(\mathbf{p}_{i,k}(j))$ is a $k \times \max(m_s, s = 1, \dots, k)$ matrix such that $c_{sw} = \tilde{v}_{sw}$ if $p_{sw} = 0$ and $c_{sw} = \tilde{u}_{sw}$ if $p_{sw} = 1$, where $\tilde{u}_{sw}, \tilde{v}_{sw}, p_{sw}$, and c_{sw} denote the (s, w)-th element of the corresponding matrix and $CHY(\mathbf{c}(\mathbf{p}_{i,k}(j)))$ is the clusterized hierarchical copula computed for the j-th o.c.c.d..*

Example 2.5.2 We consider a three-dimensional problem where the elements are grouped into two classes, the first with cardinality one and the second one composed by two elements. The dependence structure is represented by a clusterized hierarchical copula CHY in whom we compute the volume in the box $\mathbf{S} = [\mathbf{u}, \mathbf{v}]$,

[6] The coordinates of the box $\mathbf{S}$ are computed as $\mathbf{u} = col(\tilde{\mathbf{u}}')$, $\mathbf{v} = col(\tilde{\mathbf{v}}')$, where the function $col : [0, 1]^{\max\{m_s, s=1,...,k\}} \times [0, 1]^k \rightarrow [0, 1]^n$ sets orderly all the columns of a matrix into the same column.

[7] We refer to $D(i, k)$ as the number of the ways in which one can distribute the integer i into k groups taking into account the order into the groups (o.c.ds for short). If we also take into account the cardinalities of the groups, we refer to $D^c(i, k)$ as the number of the ordered compatible (for groups' cardinalities) combinatorial distributions.

where $\mathbf{u}' = [u_1, u_2, u_3]$ and $\mathbf{v}' = [v_1, v_2, v_3]$. Following the statement of the Proposition 2.5.3, the box is represented as a matrices' product, i.e.,

$$\begin{bmatrix} u_1 & u_1 \\ u_2 & u_3 \end{bmatrix} \times \begin{bmatrix} v_1 & v_1 \\ v_2 & v_3 \end{bmatrix}$$

and we generate the $\mathbf{p}_{i,2}$ matrices and the corresponding $\mathbf{c}(\mathbf{p}_{i,2})$, $i = 0, 1, 2, 3$ ones, i.e.,

$$\mathbf{p}_{0,2}(1) = \begin{pmatrix} 0 & 0 \\ 0 & 0 \end{pmatrix} \rightarrow \mathbf{c}(\mathbf{p}_{0,2}(1)) = \begin{pmatrix} v_1 & v_1 \\ v_2 & v_3 \end{pmatrix}$$

$$\mathbf{p}_{1,2}(1) = \begin{pmatrix} 1 & 0 \\ 0 & 0 \end{pmatrix} \rightarrow \mathbf{c}(\mathbf{p}_{1,2}(1)) = \begin{pmatrix} u_1 & u_1 \\ v_2 & v_3 \end{pmatrix}$$

$$\mathbf{p}_{1,2}(2) = \begin{pmatrix} 0 & 0 \\ 1 & 0 \end{pmatrix} \rightarrow \mathbf{c}(\mathbf{p}_{1,2}(2)) = \begin{pmatrix} v_1 & v_1 \\ u_2 & v_3 \end{pmatrix}$$

$$\mathbf{p}_{1,2}(3) = \begin{pmatrix} 0 & 0 \\ 0 & 1 \end{pmatrix} \rightarrow \mathbf{c}(\mathbf{p}_{1,2}(3)) = \begin{pmatrix} v_1 & v_1 \\ v_2 & u_3 \end{pmatrix}$$

$$\mathbf{p}_{2,2}(1) = \begin{pmatrix} 1 & 0 \\ 1 & 0 \end{pmatrix} \rightarrow \mathbf{c}(\mathbf{p}_{2,2}(1)) = \begin{pmatrix} u_1 & v_1 \\ u_2 & v_3 \end{pmatrix}$$

$$\mathbf{p}_{2,2}(2) = \begin{pmatrix} 1 & 0 \\ 0 & 1 \end{pmatrix} \rightarrow \mathbf{c}(\mathbf{p}_{2,2}(2)) = \begin{pmatrix} u_1 & v_1 \\ v_2 & u_3 \end{pmatrix}$$

$$\mathbf{p}_{2,2}(3) = \begin{pmatrix} 0 & 0 \\ 1 & 1 \end{pmatrix} \rightarrow \mathbf{c}(\mathbf{p}_{2,2}(3)) = \begin{pmatrix} v_1 & v_1 \\ u_2 & u_3 \end{pmatrix}$$

$$\mathbf{p}_{3,2}(1) = \begin{pmatrix} 1 & 0 \\ 1 & 1 \end{pmatrix} \rightarrow \mathbf{c}(\mathbf{p}_{3,2}(1)) = \begin{pmatrix} u_1 & v_1 \\ u_2 & u_3 \end{pmatrix},$$

where we observe that the second place in the first group is fictitious since it is only useful to write the problem in matrix form.
Finally we compose the hierarchical copulas as follows:

$$\begin{aligned} V_{CHY}(\mathbf{S}) = {} & CHY(v_1, v_2, v_3) \\ & - (CHY(u_1, v_2, v_3) + CHY(v_1, u_2, v_3) + CHY(v_1, v_2, u_3)) + \\ & + CHY(u_1, u_2, v_3) + CHY(u_1, v_2, u_3) + CHY(v_1, u_2, u_3) \\ & - CHY(u_1, u_2, u_3). \end{aligned}$$

Remark 2.5.2 We observe that $D(i, k)$ can be determined by assuming to have a number of groups equal to n, i.e., to the sum of groups' cardinalities, each one with cardinality equal to 1, so that one can take into account the order of the groups.

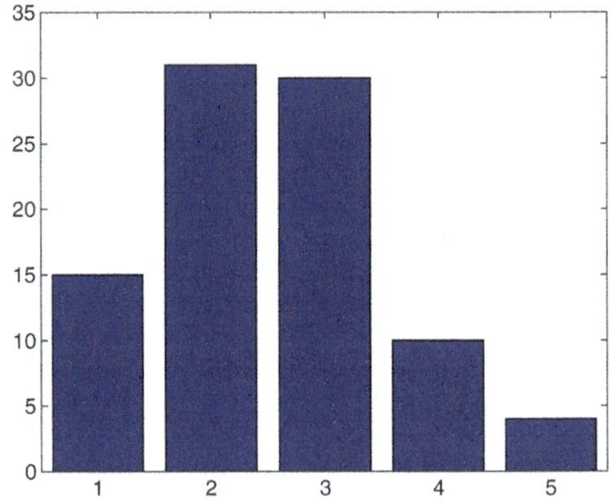

Fig. 2.13 K-means group's cardinality histogram of a sample of dimension 90

Then, computing $D(i, k) = n^i$, we must filter these o.c.d. with respect to the groups' cardinalities, reducing to the compatible ones, i.e. $D^c(i, k)$.

For example, if we have two groups each one with cardinality three, the number of o.c.ds of $i = 1$ is $D(1, 2) = D^c(1, 2) = 6$ since all the o.c.ds are compatibles with the groups' cardinalities. For all i we assume to order, in some way, the o.c.c.ds according to some order index j. This way, the second o.c.c.d. for $i = 1$ and two groups, each one with cardinality three, corresponds (for the chosen ordering criteria which is the row's order here) to

$$\mathbf{p}_{1,2}(2) = \begin{bmatrix} 0\,1\,0 \\ 0\,0\,0 \end{bmatrix}$$

where the first row corresponds to the first group and the second to the second one. We observe that we can choose any ordering criteria since the final order of the o.c.c.ds is not relevant.

2.6 Example: Homogeneous CHY-Volume Versus CR Algorithm

We consider here an insurance company, which for simplicity operates only in the life branch. The company evaluates the marginal probabilities of death of each insured but the total probability of default isn't simply assessable for an high number of policies. We face here the problem of determining the survival copula function corresponding to the probability of total survivals. We observe that for large dimensional problems like the one we are dealing with, it's necessary to reduce its complexity, and we propose here the application of a semiunsupervised clustering method. In this example we consider 90 policies. The homogeneous version of the CHY, is called CHC, and it will be theoretically discussed in Chap. 4.

The K-means clustering method based on the Euclidean metric with tolerance $\delta = 0.00031$ generates five groups with the following cardinalities:

$$clusterVector = \begin{bmatrix} 15 \; 31 \; 30 \; 10 \; 4 \end{bmatrix}$$

represented by the histogram in Fig. 2.13. These groups are assumed homogeneous with marginals equal to the centroids of the same group, which represents the vector $\mathbf{u}$ in the Corollary 4.3.2:

$$means = \begin{bmatrix} 0.0035 \; 0.00108 \; 0.00189 \; 0.0061 \; 0.0172 \end{bmatrix}$$

The dependence parameters within and between the groups are assumed equal to the EQDP in the meaning specified before:[8]

$$\alpha_{within} = \begin{bmatrix} 0.15 \; 2.5 \; 3.2 \; 0.4 \; 0.9 \end{bmatrix}$$

$$\alpha_{between} = 0.015$$

In this applications we find a survival copula function equal to 0.8324 as a sum of 90 terms with sign represented.[9]

We can fix the number of groups and the dependence parameters determined before and calculate the survival copula function for a sample where we take off an element with marginal equal to the last centroid (corresponding to the fifth group), following the order represented in *clustervector*. Working backward till a cardinality of the sample equal to 40, we can observe in Fig. 2.14 the behavior of the survival copula function with respect to the size of the sample. As expected, the survival copula function decreases for larger samples.

In order to validate our procedure, we make a comparison with the algorithm CR proposed in Cherubini and Romagnoli [6] which we consider as our Benchmark. We consider the maximum dimension which allows to implement the CR algorithm, i.e., we consider a sample of 20 heterogeneous elements (which correspond to the first 20 elements of our sample) and compare it to the clusterized homogeneous survival (CHC Survival) calculated with our procedure for several number of groups. The clusters are made with K-means and with Kohonen self-organizing map (SOM), and the distance is computed on the Euclidean metric. In Table 2.1 we report the performance of the K-means CHC procedure which clearly increases with the number of the groups.

---

[8]We assume to know the estimations of the real parameters within and between the groups which may be estimated by an econometric methodology, such as the maximum likelihood method, possibly simplifying the problem by considering the limiting distributions proposed for Archimedean copulas with granularity adjustment, and then compensated to account for the homogeneous approximation as suggested in Definition 2.4.2.

[9]We must recall the representation of the survival copula function proposed in Corollary 4.3.2.

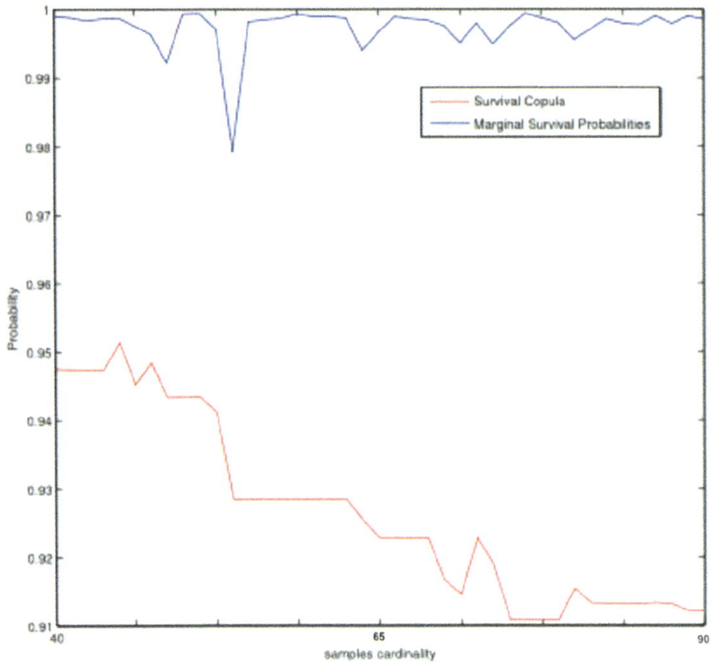

Fig. 2.14 The CHC survival copula of a sample clusterized in five groups for an increasing dimension of the sample

Table 2.1 K-means CHC Survival copula vs CR Survival for several number of groups. Clayton copula with $\alpha_{within} = 0.15$ and $\alpha_{between} = 0.015$ computed on a heterogeneous sample of dimension 20

Number of groups	K-means CHC Survival	Error
2	0.9578	0.0034
4	0.9563	0.0018
5	0.9560	0.0013
10	0.9552	0.0005

In Tables 2.2 and 2.3, we investigate the performance of the CHC procedure on a randomized data set. We start from the same data set used in Table 2.1 randomized by a white Gaussian noise and by a colored Gaussian noise, i.e., a white Gaussian noise affecting only few marginals. The clusters are made by K-means based on the Euclidean distance. We compare the CR Survival with the K-means CHC Survival, computed on a randomized data by a Monte Carlo procedure. On a random sample, we compute the K-means CHC Survival for a maximum number of groups equal to 5 since K-means can't preserve a greater number of clusters during the iterations, reducing it when one of the clusters becomes empty.

Table 2.2 K-means CHC Survival copula vs CR Survival for several number of groups. Clayton copula with $\alpha_{within} = 0.15$ and $\alpha_{between} = 0.015$ computed on a white randomized heterogeneous sample of dimension 20

Number of groups	K-means CHC Survival/white random	Error
2	0.9577	0.0039
4	0.9569	0.0018
5	0.9562	0.0013

Table 2.3 K-means CHC Survival copula vs CR Survival for several numbers of groups. Clayton copula with $\alpha_{within} = 0.15$ and $\alpha_{between} = 0.015$ computed on a colored randomized heterogeneous sample of dimension 20

Number of groups	K-means CHC Survival/colored random	Error
2	0.9264	0.0265
4	0.9235	0.0315
5	0.9223	0.033

Table 2.4 SOM CHC Survival copula vs CR Survival in several scenarios. Clayton copula with $\alpha_{within} = 0.15$ and $\alpha_{between} = 0.015$

Scenario	SOM CHC Survival	Error
Deterministic	0.9568	0.0009
White random	0.9569	0.002
Colored random	0.9236	0.037

We can observe that for an higher number of groups, the error decreases for the white randomization but increases for the colored one. In line with Francois et al. [16], we assert that the K-means CHC approach based on the Euclidean metric can be a good choice in the presence of white Gaussian noise, but the performance is not so good if we assume a colored randomization of data.

In Table 2.4 we compare the CR Survival with the Kohonen SOM CHC Survival computed on the data set of dimension 20, allowing also for white and colored randomization. Kohonen SOM groups the data on a map with six neurons, and, yet after 200 iterations, the grid tends to approximate the data distribution. In the random data experiments, we compute SOM CHC Survival by a Monte Carlo procedure.

If we compare Table 2.4 with the previous tables relative to K-means applications, we observe that SOM with a small number of nodes behaves similarly to K-means, while, as it has been pointed in Kohonen and Hondela [21], larger self-organizing maps rearrange data in a way that is fundamentally topological in character.

Table 2.5 CPU computation time in seconds. K-means CHC probability distribution of the left queue (till 5 percent) of a variable of dimension $n = 100$

N. of groups	Numerosity of the groups	Cpu time
5	20	0.65
10	10	9
20	5	227

Table 2.6 CPU computation time in seconds. K-means CHC probability distribution of the left queue (till 10 percent) of a variable of dimension $n = 60$

N. of groups	Numerosity of the groups	Cpu time
5	12	0.9
10	6	19.75
15	4	177

2.6.1 Scalability of the Homogeneous CHY-Based Algorithm

An interesting question concerns the scalability of the algorithm, i.e., the maximum dimension to which our techniques can be extended. In K-means CHC algorithm (standing for Homogeneous CHY), the scalability must be analyzed with respect to the number of groups and the total dimension of the problem. The MatLab™ algorithm was lunched on a Dell Dimension 2,8 GHz Intel Core i7 quad-core, 16 GB RAM. Table 2.5 reports the computing time for the first 5 percent of the probability distribution of a variable of dimension $n = 100$ for different numbers of groups. Table 2.6 reports the computing time for the first 10 percent of the probability distribution of a variable of dimension $n = 60$ for different numbers of groups. We can observe that the number of groups is the critical variable; the CPU time increases exponentially with the number of group till 15, for the next divisor, i.e., 20, problems of memory stack become unmanageable.

References

1. Barbe, P., Genest, C., Ghoudi, K., & Rémillard, B. (1996). On Kendall's process. *Journal of Multivariate Analysis, 58*, 197–229.
2. Bedford, T., & Cooke, R. M. (2002). Vines: A new graphical model for dependent random variables. *Annals of Statistics, 30*(4), 1031–1068.
3. Bernardi, E., & Romagnoli, S. (2011). Computing the volume of an high-dimensional semi-unsupervised hierarchical copula. *International Journal of Computer Mathematics, 88*(12), 2591–2607.
4. Bernardi, E., & Romagnoli, S. (2012). Limiting loss distribution on a hierarchical copula-based model. *International Review of Applied Financial Issues and Economics, 4*(2), 126–145.
5. Bernardi, E., & Romagnoli, S. (2013). A clusterized copula-based probability distribution of a counting variable for high-dimensional problems. *The Journal of Credit Risk, 9*(2), 3–26.

6. Cherubini, U., & Romagnoli, (2009). Computing copula volume in n-dimensions. *Applied Mathematical Finance, 16*(4), 307–314.
7. Choros, B., Härdle, W., & Okhrin, O. (2013). Valuation of collateralized debt obligations with hierarchical Archimedean copulae. *Journal of Empirical Finance, 24*, 42–62.
8. Cifuentes, A., & O'Connor, G. (1996). The binomial expansion method applied to CBO/CLO analysis. *Moody's Special Report*, December.
9. Cifuentes, A., Efrat, I., Glouck, J., & Murphy, E. (1999). Buying and selling credit risk: A perspective on credit-linked obligations. In J. Gregory (Ed.), *Credit Derivatives* (pp. 112–123). London: Risk Book.
10. Czado, C. (2010). Pair-copula constructions of multivariate copulas. In P. Jaworski, F. Durante, W. Härdle, & T. Rychlik (Eds.), *Copula Theory and Its Applications*. Berlin: Springer.
11. dos Anjos, U., & Kolev, N. (2005). An application of Kendall distributions. *Review of Business and Economic Research, 1*, 95–102.
12. Duffie, D., & Gârleanu, N. (2001). Risk and valuation of collateralized debt obligations. *Financial Analyst's Journal, 57*(1), 41–59.
13. Embrechts, P., Lindskog, F., & McNeil, A. (2003). Modelling dependence with copulas and applications to risk management. In S. Rachev (Ed.), *Handbook of Heavy Tailed Distributions in Finance* (pp. 329–384). Amsterdam/Boston: Elsevier
14. Fang, K. T., Kots, S., & Ng, W. (1990). *Symmetric Multivariate and Related Distributions*. London: Chapman & Hall.
15. Féron, R. (1956). Sur les tableaux de corrélation dont les marges sont donées, cas de l'espace à trois dimensions. *Publ. Inst. Statist. Univ. Paris, 5*, 3–12.
16. Francois, D., Wertz, V., & Verleysen, M. (2005). Non-Euclidean metrics for similarity search in noisy data sets. In *Proceedings of European Symposium on Artificial Neural Networks*, ESANN'2005.
17. Hering, C., Hofert, M., Mai, J. F., & Scherer, M. (2010). Constructing hierarchical Archimedean copulas with Lèvy subordinators. *Journal of Multivariate Analysis, 101*, 1428–1433
18. Hofert, M. (2008). Sampling Archimedean copulas. *Computational Statistics and Data Analysis, 52*(12), 5163–5174.
19. Imlahi, L., Ezzerg, M., & Chakak, A. (1999). Estimación de la curva mediana de una cópula $C(x_1, \ldots, x_n)$. *Rev. R. Acad. Cien. Exact. Fis. Nat, 93*(2), 241–250.
20. Joe, H. (1997). *Multivariate Models and Dependence Concepts*. London: Chapman & Hall.
21. Kohonen, T., & Hondela, T. (2007). Kohonen network. *Scholarpedia, 2*, 1568.
22. Kurowicka, D., & Joe, H. (2011). *Dependence Modeling. Vine Copulae Handbook*. Singapore/Hackensack: World Scientific.
23. Luo, X., & Shevchenko, P. V. (2010). The t copula with multiple parameters of degrees of freedom: bivariate characteristics and application to risk management. *Quantitative Finance, 10*(9), 1039–1054.
24. McNeil, A. J., & Nešlehová, J. (2009). Multivariate Archimedean copulas, d-monotone functions and l_1-norm symmetric distributions. *Annals of Statistics, 37*(5B), 3059–3097.
25. Nelsen, R. N. (2006). *Introduction to Copulas* (2nd ed.). Heidelberg: Springer.
26. Okhrin, O., Okhrin, Y., & Schmid, W. (2013a). Determining the structure and estimation of hierarchical Archimedean copulas. *Journal of Econometrics, 173*(2), 189–204.
27. Okhrin, O., Okhrin, Y., & Schmid, W. (2013b). Properties of hierarchical Archimedean copulas. *Statistics and Risk Modeling, 30*(1), 21–53.
28. Pickands, J. (1981). Multivariate extreme value distributions. *Bulletin of the International Statistical Institute, 49*, 859–878.
29. Savu, C., & Trede, M. (2010). Hierarchical Archimedean Copulas. *Quantitative Finance, 10*(3), 295–304.
30. Schorin, C., & Weinrich, S. (1998). Collateralized debt obligation handbook. *Working paper*, Morgan Stanley Dean Witter.
31. Sklar, A. (1959). Fonctions de repartition à n dimensions et leurs marges. *Publication Inst. Statist. Univ. Paris, 8*, 229–231.

Chapter 3
Combinatorics and Random Matrices: A Brief Review

Abstract The purpose of this chapter is to briefly introduce and review some fundamental concepts in counting and combinatorics which will be used in the subsequent chapters extensively.

The goal of this chapter is to present some fundamental theorems and the necessary tools and ideas in counting and combinatorics which would help the reader navigate smoothly through the rest of the text. An advanced reader can safely skip this chapter. The references for this chapter, while being well-known, are mainly presented following Strirzaker [3], Shiryaev [2] and Ross [1].

3.1 Combinatoric Distribution of a Random Event

Many interesting problems in probability can be solved by counting the number of outcomes in an event. Such counting often turns out to also be useful in more general contexts. This chapter sets out some simple methods of dealing with the commonest counting problems which we will apply in this book. See, e.g., Remark 4.3.1 and the discussion following Proposition 4.3.1.

The basic principles are pleasingly easy and are enumerated below.

1. **The basic principal of counting**: Suppose that two experiments to be performed. Then if experiment 1 can result in any one of m possible outcomes and if for each outcome of experiment 1, there are n possible outcomes of experiment 2, and then together there are mn possible outcomes.
2. **The generalized basic principle of counting**: If r experiments that are to be performed are such that the first one may result in any of n_1 possible outcomes, if for these n_1 possible outcomes there are n_2 possible outcomes of the second experiment, if for each of the possible outcomes of the first two experiments there are n_3 possible outcomes for the third experiment, and if ..., then there is a total of $n_1 \cdot n_2 \cdots n_r$ possible outcomes of the r experiments.

E. Bernardi, S. Romagnoli, *Counting Statistics for Dependent Random Events*,
https://doi.org/10.1007/978-3-030-64250-1_3

Real problems involve, for example, catching different varieties of fish, drawing various balls from a number of urns, and dealing hands at numerous types of card games. In the standard terminology for such problems, we say that a number n (say) of objects or things are to be divided or distributed into r classes or groups. The number of ways in which this distribution can take place depends on whether

1. The objects can be distinguished or not.
2. The classes can be distinguished or not.
3. The order of objects in a class is relevant or not.
4. The order of classes is relevant or not.
5. The objects can be used more than once or not at all.
6. Empty classes are allowed or not.

3.1.1 Permutations: Ordered Selection

How many different ordered arrangements of letters a, b, and c are possible? By direct enumeration we see that there are 6, namely, abc, acb, bac, bca, cab, $andcba$. Each arrangement is known as a permutation. There are 6 possible permutations of a set of 3 objects. This result could also have been obtained observing that the first object in the permutation can be any of the 3, the second object in the permutation can be chosen from any of the remaining 2, and the third object in the permutation is then chosen from the remaining 1. Thus there are $3 \cdot 2 \cdot 1$ possible permutations. Suppose now that we have n objects. Reasoning similar to that we have just used for the 3 letters shows that we have

$$n(n-1)(n-2)\cdots 3 \cdot 2 \cdot 1 = n!$$

permutations. We now state certain important formulas on computing the number of permutations in different situations which might come in handy later in the text.

Theorem 3.1.1 *Given n distinct symbols, the number of distinct permutations (without repetition) of length $r \leq n$ is*

$$n(n-1)\cdots(n-r+1) = \frac{n!}{(n-r)!}$$

Theorem 3.1.2 *Given n distinct symbols which may be repeated any number of times, the number of permutations of length r is n^r.*

Theorem 3.1.3 *Given $n = \sum_{i=1}^{r} n_i$ symbols of r distinct types, where n_i are of type i and are otherwise indistinguishable, the number of permutations (without repetition) of all n symbols is*

$$M_n(n_1, \ldots, n_r) = \frac{n!}{\prod_{i=1}^{r} n_i!}$$

3.1.2 Combinations: Unordered Selection

We are often interested in determining the number of different groups of r objects that could be formed from a total of n objects.

In general, as $n(n-1)\cdots(n-r+1)$ represents the number of different ways that a group of r items could be selected from a group of n items when the order of the selection is relevant and as each group of r items will be counted $r!$ times in this count, it follows that the number of different groups of r items that could be formed from a set of n items is

$$\frac{n(n-1)\cdots(n-r+1)}{r!} = \frac{n!}{(n-r)!r!}$$

Now we state a bunch of results on combinations which are used extensively is this text as theorems. Note that Theorem 3.1.4 summarizes the discussion above.

Theorem 3.1.4 *The number of ways of choosing a set of r symbols from a set of n distinct symbols without repetition is*

$$\binom{n}{r} = \frac{n!}{r!(n-r)!}$$

A set of n distinct items is to be divided into r distinct groups of respective sizes $n_1, n_2, \ldots n_r$ where $\sum_{i=1}^{r} n_i = n$. How many different ways are possible? To answer this, we note that there are $\binom{n}{n_1}$ possible choices for the first group; for each choice of the first group, there are $\binom{n-n_1}{n_2}$ choices for the second group; for each choice of the first two groups, there are $\binom{n-n_1-n_2}{n_3}$ possible choices for the third group; and so on and so forth. Hence it follows by the generalized version of the basic counting principle that there are

$$\binom{n}{n_1}\binom{n-n_1}{n_2}\binom{n-n_1-n_2}{n_3}\cdots\binom{n-n_1-n_2-\cdots-n_{r-1}}{n_r}$$
$$= \frac{n!}{(n-n_1)!n_1!}\frac{(n-n_1)!}{(n-n_1-n_2)!n_2!}\cdots\frac{(n-n_1-n_2-\cdots-n_{r-1})!}{0!n_r!}$$

possible divisions. This reasoning gives rise to the following well-known results (see, e.g., Shiryaev 2)

Theorem 3.1.5 *The number of ways of dividing n distinct objects into r distinct groups of sizes $n_1, n_2, \ldots, n_r$, where $\sum_{i=1}^{r} n_i = n$ is $\frac{n!}{\prod_{i=1}^{r} n_i!}$.*

Theorem 3.1.6 *Out of $n = \sum_{i=1}^{r} n_i$ symbols (where the ith distinct set contains n_i indistinguishable symbols), we can select $\prod_{i=1}^{r}(n_i+1)-1$ combinations.*

Theorem 3.1.7 *Given a set of n distinct symbols that may be repeated any number of times, the number of ways of choosing a set of size r is* $\binom{n+r-1}{r}$

We end this section with a couple of very useful results which might come handy later

Theorem 3.1.8 *There are* $\binom{n-1}{r-1}$ *distinct positive integer-valued vectors* $(x_1, x_2, \ldots, x_r)$ *satisfying*

$$x_1 + x_2 + \cdots + x_r = n \; x_i > 0, i = 1, \ldots r$$

Theorem 3.1.9 *There are* $\binom{n+r-1}{r-1}$ *distinct nonnegative integer-valued vectors* $(x_1, x_2, \ldots, x_r)$ *satisfying*

$$x_1 + x_2 + \cdots + x_r = n$$

3.1.3 The Hardy-Ramanujan Asymptotic Partition Formula

Some of the algorithmic and numerical complexities arising later in our treatment of certain random matrices have a direct theoretical origin in the following classical result of Hardy and Ramanujan.

In number theory and combinatorics, a partition of a positive integer n, also called an integer partition, is a way of writing n as a sum of positive integers. Two sums that differ only in the order of their summands are considered the same partition. (If order matters, the sum becomes a composition.) For example, 4 can be partitioned in five distinct ways:

4
$3 + 1$
$2 + 2$
$2 + 1 + 1$
$1 + 1 + 1 + 1$

The order-dependent composition $1 + 3$ is the same partition as $3 + 1$, while the two distinct compositions $1 + 2 + 1$ and $1 + 1 + 2$ represent the same partition $2 + 1 + 1$.

A summand in a partition is also called a part. The number of partitions of n is given by the partition function $p(n)$. So $p(4) = 5$.

Partitions can be graphically visualized with Young diagrams or Ferrers diagrams. They occur in a number of branches of mathematics and physics, including the study of symmetric polynomials, the symmetric group, and in group representation theory in general.

No closed-form expression for the partition function is known, but it has both asymptotic expansions that accurately approximate it and recurrence relations by which it can be calculated exactly. It grows as an exponential function of the square root of its argument.

Theorem 3.1.10 *The Hardy-Ramanujan Asymptotic partition formula: for a positive integer n, let p(n) denote the number of unordered partitions of n, that is, unordered sequence of positive integers which sum to n; then the value of p(n) is asymptotically given by*

$$p(n) \approx \frac{1}{4n\sqrt{3}} e^{\tau\sqrt{\frac{n}{6}}}$$

with a certain fixed constant τ.

3.1.4 The Combinatorial Problem in CHY-Volume Computation

To compute the CHY-volume, as described in Proposition 2.5.1, we must solve an interesting combinatorial problem. In fact we are concerned with the computing of the combinatoric distributions of an integer into the groups already determined by a clustering procedure. These scenarios must be compatible with the complexity of the groups; hence we talk about *compatible combinatoric distributions* (c.c.ds for short). We must consider the c.c.ds for all the integers from zero to the dimension of the problem. The following MatLab™ code explains the procedure.

```
plist = zeros(0,n);
for i = 0:m
temp = partitions(total_sum - i*c, ...
        candidate_set(1:(end-1)), ...
max_count(1:(end-1)),fixed_count-i);
plist = [plist;[temp,repmat(i,size(temp,1),1)]];
end
```

3.1.5 Testing Compatibility with the Groups

Moving beyond the purely combinatorial framework, we now select the c.ds consistent with the groups' cardinalities generated by a clustering methods as K-means. The following MatLab™ code explains the procedure.

```
% This function tests the compatibility of the c.c.ds in MPerm
% with the groups' cardinalities.

function[SizeThreshold]= SizeThreshold(MPerm,p)

% If the number of ones in the group is bigger than its
```

```
% cardinality, the row is set to zero and then erased.

nc = size(MPerm,2);
for j=1:size(MPerm,1)
for i=1:nc
if (MPerm(j,i) > p(1,i))
MPerm(j,:)=0;
else
MPerm(j,i) = MPerm(j,i);
end
end
end
SizeThreshold=MPerm;
```

3.2 Random Matrices

In this section we will give a very brief introduction to the most important concepts in Random matrices which will be used extensively in the subsequent chapters. In probability theory and mathematical physics, a random matrix is a matrix-valued random variable, that is, a matrix in which some or all of the entries are random variables. Many important properties of physical systems can be represented mathematically as matrix problems. For example, in this book arrival matrices which represent the arrival policy of a random event are random matrices and will be dealt with extensively in the coming chapters.

We begin by first outlining the most important class of random matrices used in applications such as physics, mathematical statistics and numerical analysis, number theory, theoretical neuroscience, and optimal control.

3.2.1 *Gaussian Ensembles*

The **Gaussian unitary ensemble GUE**(n) is described by the Gaussian measure with density

$$\frac{1}{Z_{\mathrm{GUE}(n)}} e^{\frac{-n}{2} \operatorname{tr} H^2}$$

on the space of $n \times n$ Hermitian matrices $H = (H_{ij})_{1 \leq i,j \leq n}$. Here $Z_{\mathrm{GUE}(n)} = 2^{n/2}\pi^{n^2/2}$ is a normalization constant, chosen so that the integral of the density is equal to one. The term unitary refers to the fact that the distribution is invariant under unitary conjugation. The Gaussian unitary ensemble models Hamiltonians lacking time-reversal symmetry.

The **Gaussian orthogonal ensemble GOE(n)** is described by the Gaussian measure with density

$$\frac{1}{Z_{\mathrm{GOE}(n)}} e^{\frac{-n}{4} \operatorname{tr} H^2}$$

on the space of $n \times n$ real symmetric matrices $H = (H_{ij})_{1 \le i,j \le n}$. Its distribution is invariant under orthogonal conjugation, and it models Hamiltonians with time-reversal symmetry.

The **Gaussian symplectic ensemble GSE(n)** is described by the Gaussian measure with density

$$\frac{1}{Z_{\mathrm{GSE}(n)}} e^{-n \operatorname{tr} H^2}$$

on the space of $n \times n$ Hermitian quaternionic matrices, e.g., symmetric square matrices composed of quaternions, $H = (H_{ij})_{1 \le i,j \le n}$. Its distribution is invariant under conjugation by the symplectic group, and it models Hamiltonians with time-reversal symmetry but no rotational symmetry.

The joint probability density function for the eigenvalues $\lambda_1, \lambda_2, \ldots, \lambda_n$ of GUE/GOE/GSE random matrices defined above is given by

$$\frac{1}{Z_{\beta,n}} \prod_{k=1}^{n} e^{\frac{-\beta n}{4} \lambda_k^2} \prod_{i<j} |\lambda_j - \lambda_i|^{\beta}$$

where the Dyson index $\beta = 1$ for GOE, $\beta = 2$ for GUE, and $\beta = 4$ for GSE counts the number of real components per matrix element; $Z_{\beta,n}$ is a normalization constant which can be explicitly computed; see the Selberg integral in Livan, G. - Novaes, M. - Vivo,P. (2018). The Gaussian factor $\prod_{k=1}^{n} e^{-\frac{\beta n}{4} \lambda_k^2}$ kills any configuration of eigenvalues $\{\lambda\}$ where some λ_j's are big (far from zero in absolute value): the eigenvalues do not like to stay too far from the origin. On the other hand, the term $\prod_{i<j} |\lambda_j - \lambda_i|^{\beta}$ kills the configuration where the eigenvalues get "too close" to each other.

The repulsion factor $\prod_{i<j} |\lambda_j - \lambda_i|^{\beta}$ has another effect: it makes the eigenvalues strongly nonindependent. Every eigenvalue feels the presence of all the others, and the joint probability density function does not factorize at all. Hence, the classical tools for independent random variables are of little use here. This interplay between confinement and repulsion is the physical mechanism at the heart of many results in Random matrix theory.

3.2.2 An Illustrative Example of a 2 × 2 Random Matrix

Consider a 2 × 2 GOE matrix $H_s = \begin{bmatrix} x_1 & x_3 \\ x_3 & x_2 \end{bmatrix}$ with $x_1, x_2 \sim N(0, 1)$ and $x_3 \sim N(0, 1/2)$. What is the pdf of $p(s)$ of the spacing $s = \lambda_2 - \lambda_1$ between its two eigenvalues ($\lambda_2 > \lambda_1$)?

The two eigenvalues are random variables, given in terms of the entries by the roots of the characteristic polynomial

$$\lambda^2 - \mathrm{tr}(H_s)\lambda + \det(H_s)$$

and therefore $\lambda_{1,2} = \left(x_1 + x_2 \pm \sqrt{(x_1 - x_2)^2 + 4x_3^2}\right)/2$ and $s = \sqrt{(x_1 - x_2)^2 + 4x_3^2}$.
By definition we have

$$p(s) = \int_{-\infty}^{\infty} dx_1 dx_2 dx_3 \frac{e^{-\frac{1}{2}x_1^2}}{\sqrt{2\pi}} \frac{e^{-\frac{1}{2}x_2^2}}{\sqrt{2\pi}} \frac{e^{-x_3^2}}{\sqrt{\pi}} \delta\left(s - \sqrt{(x_1 - x_2)^2 + 4x_3^2}\right) \tag{3.1}$$

Changing the variables as

$$\begin{cases} x_1 - x_2 = r\cos\theta \\ 2x_3 = r\sin\theta \\ x_1 + x_2 = \varphi \end{cases}$$

we get

$$\begin{cases} x_1 = \frac{r\cos\theta+\varphi}{2} \\ x_2 = \frac{\varphi - r\cos\theta}{2} \\ x_3 = \frac{r\sin\theta}{2} \end{cases}$$

and computing the corresponding Jacobian

$$J = \det\begin{bmatrix} \frac{\partial x_1}{\partial r} & \frac{\partial x_1}{\partial \theta} & \frac{\partial x_1}{\partial \varphi} \\ \frac{\partial x_2}{\partial r} & \frac{\partial x_2}{\partial \theta} & \frac{\partial x_2}{\partial \varphi} \\ \frac{\partial x_3}{\partial r} & \frac{\partial x_3}{\partial \theta} & \frac{\partial x_3}{\partial \varphi} \end{bmatrix} = \det\begin{bmatrix} \cos\theta/2 & -r\sin\theta/2 & 1/2 \\ -\cos\theta/2 & r\sin\theta/2 & 1/2 \\ \sin\theta/2 & r\cos\theta/2 & 0 \end{bmatrix} = -r/4$$

one obtains

$$p(s) = \int_0^{\infty} dr\; r\delta(s-r) \int_0^{2\pi} d\theta \int_{-\infty}^{\infty} d\varphi e^{-\frac{1}{2}\left[\left(\frac{r\cos\theta+\varphi}{2}\right)^2+\left(\frac{-r\cos\theta+\varphi}{2}\right)^2+\frac{r^2\sin^2\theta}{2}\right]}$$
$$= \frac{\sqrt{4\pi}s}{8\pi^{3/2}} \int_0^{2\pi} d\theta e^{-\frac{1}{2}\left[\frac{s^2\cos^2\theta}{2}+\frac{s^2\sin^2\theta}{2}\right]} = \frac{s}{2}e^{-s^2/4} \tag{3.2}$$

Note that we used $\cos^2\theta+\sin^2\theta = 1$ to achieve this very simple result; however, we could only use this massive simplification because the variance of the off-diagonal element was $1/2$ of the variance of the diagonal elements. It is also not very hard to verify that the pdf is correctly normalized, i.e., $\int_0^{\infty} dsp(s) = 1$.

3.2.3 Singular Values of Rectangular Matrices

Let A be a matrix and A^* be its transpose if A is real or the conjugate of the transpose if A is complex. Then σ is a singular value of A and u and v its corresponding singular vectors if $Av = \sigma u$ and $A^*u = \sigma v$. Notice that $A^*A = \sigma A^*u = \sigma^2 v$ and $AA^*u = \sigma Av = \sigma^2 u$ so that σ^2 is an eigenvalue of A^*A and an eigenvalue of AA^*. One of the uses of singular values is in the singular value decomposition where A can be written as $A = U\Sigma V^*$ where Σ is a matrix (only square if A is) with the singular values on the diagonal and zeros elsewhere, i.e., σ_{ii} is a singular value.

3.2.4 Marchenko-Pastur Distribution

In the mathematical theory of random matrices, the Marchenko-Pastur distribution, or Marchenko-Pastur law, describes the asymptotic behavior of singular values of large rectangular random matrices. The theorem is named after Ukrainian mathematicians Vladimir Marchenko and Leonid Pastur who proved this result in 1967.

Let

$$X = (X_1\, X_2, \ldots, X_n) \in \mathbb{R}^{p\times n}$$

where X_{ij} are i.i.d, $E(X_{ij}) = 0$, $E(X_{ij}^2) = 1$ and $p = p(n)$.

Define

$$S_n = \frac{1}{n}XX^T \in \mathbb{R}^{p\times p},$$

and let

$$\lambda_1 \leq \lambda_2 \cdots \leq \lambda_p$$

denote the eigenvalue of the matrix S_n. Define the random spectral measure by

$$\mu_n = \frac{1}{p}\sum_{i=1}^{p} \mathbf{1}_{\{\lambda_i\}}$$

We are now ready to state the Marchenko-Pastur law

Theorem 3.2.1 *Let S_n, μ_n be as above. Assume that $p/n \to y \in (0, 1]$ as $n \to \infty$. Then we have*

$$\mu_n(., \omega) \Rightarrow \mu \ a.s$$

where μ is a deterministic measure whose density is given by

$$\frac{d\mu}{dx} = \frac{1}{2\pi xy}\sqrt{(b-x)(x-a)}1_{(a\leq x\leq b)} \tag{3.3}$$

and here a and b are functions of y given by

$$a(y) = (1-\sqrt{y})^2, b(y) = (1+\sqrt{y})^2$$

The weak limiting distribution μ in the theorem is called the Marchenko-Pastur distribution.

Remark 3.2.1 If $y > 1$, and then since the rank$(S) = p \wedge n$, we will roughly have $n(y-1)$ zero eigenvalues. Since $\mu_n = \frac{1}{p}\sum_{i=1}^{p} \mathbf{1}_{\{\lambda_i\}}$, we see that there will be a mass of $(1 - y^{-1})$ at 0 in the limiting measure. Since the nonzero eigenvalues of XX^T and $X^T X$ are the same, we can say that in this limiting case the limiting distribution is

$$(1 + y^{-1})\mathbf{1}_{\{0\}} + \mu$$

where μ satisfies (3.3)

Remark 3.2.2 Observe that if $y = 1$, then $a = 0, b = 4$, and thus

$$\frac{d\mu}{dx} = \frac{1}{2\pi x}\sqrt{(4-x)(x)}1_{(0\leq x\leq 4)}$$

In this case μ is image of the semicircle distribution under the mapping $x \mapsto x^2$

3.2.5 *The Distorted Combinatoric Distributions*

To anticipate the argument of Chap. 5, we can imagine to have a system where the arrival of random event are controlled by idiosyncratic factors and are influenced by the dependence structure of the network at all. This way the set of combinatoric distributions will be composed by several scenarios, everyone assigned to different probability. The equal probability assumption cannot be supported in this model. In order to generate the most probable scenarios in this kind of model, we refer to a kind of Monte Carlo approach based on a generator which is a random matrix. We called it *the set of distorted combinatoric distribution*. The set of distorted distributions is recovered by a sampling procedure where both the marginals and the dependence structure affect the arrival of the random event. The following MatLab™ code explains the procedure.

```
%This function computes the combinatoric distribution distorted by the
%dependence structure among variables. The parameters rho in Dmatrix
% correspond to the dependence parameter of the hierarchical copula

%Here we sample from a multivariate normal distribution.
% Mean is the row vector of the mean.
% Dmatrix is the dispersion matrix of the multivariate normal distribution.
% vectp is a row vector of the probability that a Bernoulli variable gets 1
% i.e. it is the default probability
% dim is the number of trials

function[RM]= RandomMatrix(Dmatrix,vectp,dim)
mean=0*ones(1,size(Dmatrix,1))
Z=mvnrnd(mean,Dmatrix,dim)
U=normcdf(Z)
for i=1:size(vectp,2)
    X(:,i)=binoinv(U(:,i),1,vectp(i))
end
            RM=[X(:,:)]
```

References

1. Ross, S. (1998). *A First Course in Probability* (5th ed.). London: Prentice Hall.
2. Shiryaev, A. (2016). *Probability-1*. New York: Springer.
3. Strirzaker, D. (2003). *Elementary Probability* (2nd ed.). Cambridge: Cambridge University Press.

Part II
Mixing the Ingredients: A Recipe for a New Aggregation Algorithm

Abstract In Part II the ingredients presented in Part I are arranged in such a way that the resulting new aggregation algorithm is ready for a detailed analysis. Several applications based on real data are described eventually, and the performance comparison of the new algorithm to alternative approaches already in the literature is considered and further analyzed.

Chapter 4
Counting a Random Event: Traditional Approach and New Perspectives

Abstract The goal of this chapter is to recover the distribution function of a counting random variable representing a countable event defined on a set of multidimensional variables, whose dependence structure is known. After a review of the existing main contributions in counting statistics, we will introduce our new approach to the problem.

Our interest is here to recover the distribution function of a counting random variable representing a countable event defined on a set of multidimensional variables, whose dependence structure is known. This event could be in fact the default or the loss of a financial product or the death of an insured person or any other random event that affects one or more dependent random variables. With this intention in mind, we review at first the historical main contributions in counting statistics in order to eventually introduce our new approach to the problem.

4.1 Counting Variables: Fundamentals in Literature

We talk about a counting variable when we consider a random variable that counts the events occurred in a fixed interval of time and/or space. In a seminal work, Poisson [19] focused on a particular random variable that counts the number of discrete occurrences (sometimes called *events* or *arrivals*) which take place during a time interval of a given length. The basic assumption is that the events take place with a known average rate and independently of the time since the last event. The probability distribution of this random variable is known as the *Poisson distribution*.

Definition 4.1.1 (Poisson distribution) A discrete random variable N is said to have a Poisson distribution with parameter $\lambda > 0$, if for $k = 0, 1, \ldots$ the probabilities of N are given by

$$\mathbb{P}(N = k) = \frac{\lambda^k e^{-\lambda}}{k!}.$$

E. Bernardi, S. Romagnoli, *Counting Statistics for Dependent Random Events*,
https://doi.org/10.1007/978-3-030-64250-1_4

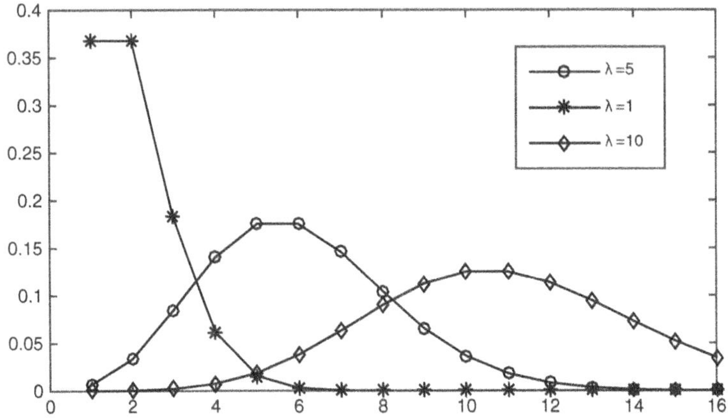

Fig. 4.1 Poisson probability distribution for different values of the Poisson parameter λ

The expectation and the variance of X are both equal to $\lambda \in \mathbb{R}^+$.

In Fig. 4.1 the Poisson probability distribution is depicted for different values of the Poisson parameter λ.

The Poisson distribution is also called the *law of small numbers* because it represents the probability distribution of the number of instances of an event which happens rarely but with very many opportunities to occur. Sometimes the Poisson distribution is called *Bortkiewicz distribution* for the contribution to the literature on counting variables given by Bortkiewicz [3].

The Poisson distribution can be recovered as a limiting case of a binomial distribution, assuming that the number of trials goes to infinity and the expected number of successes remain fixed. This implies that the Poisson distribution can be used as an approximation of the binomial one if n is sufficiently large and p is sufficiently small, i.e., $F_{Bin}(k; n, p) \approx F_{Poi}(k; \lambda = np)$. We use to say that the Poisson distribution is a good approximation of the binomial distribution if $n \geq 20$ and $p \leq 0.05$ and an excellent approximation if $n \geq 100$ and $np \leq 10$. Moreover, for sufficiently large values of λ, i.e., for $\lambda > 1000$, the normal distribution with mean and variance equal to λ is an excellent approximation to the Poisson distribution, $F_{Poi}(k; \lambda) \approx F_{Norm}(k; \lambda, \lambda)$. Anyway for smaller λ, the normal approximation requires one to perform an appropriate continuity correction on k.

Another important property of the Poisson distribution concerns the distribution of the inter-arrival times; if for every $t > 0$ the number of arrivals in $[0, t]$ follows a Poisson distribution with mean λt, then the sequence of inter-arrival times are i.i.d. exponential random variables with mean λ^{-1}.

Applications of the Poisson distribution can be found in many fields related to the counting of the occurrences of random events, as, for instance, in telecommunications to represent telephone calls arriving in a system, in astronomy to represent photons arriving at a telescope, in biology for the number of mutations on a strand of DNA per unit length, in management for customers arriving at a counter or call center, in civil engineering to represent cars arriving at a traffic light, in finance and insurance for the number of losses/claims occurring in a given period of time, in earthquake seismology where an asymptotic Poisson model of seismic risk for large earthquakes has been proposed in Lomnitz [16], and in radioactivity to represent the decay of a radioactive nucleus.

4.1.1 Generalized Poisson Distribution

The generalized Poisson distribution assumes that the occurrence of an event does not remain constant and may change with time and/or previous occurrences, generating different values of the parameter λ in the Poisson distribution. This assumption could be valid in the case of a population whose individuals have different degrees of accident rates; members of the population may not be equally exposed to risk and therefore the propensity of the given population to a particular random event, in counting whose happenings we are interested, may vary from year to year.

The Poisson distribution has been considered in literature in a number of hypothesis, in particular with varying values of the parameters. Greenwood and Yule [11] assumed that λ in the Poisson distribution followed a gamma distribution; this assumption led them to the *negative-binomial distribution*. This distribution models the number of failures in a sequence of independent and identically distributed Bernoulli trials before a given number of successes (denoted by the parameter r) occurs. The negative-binomial distribution allows the mean and variance to be different, unlike the Poisson; its variance converges to the Poisson one in the limit $r \to +\infty$. For this reason this kind of distribution is appropriate where events have positively correlated occurrences, causing a larger variance than if the occurrences were independent. This distribution is also called *Pascal distribution* if r is an integer, while it is called *Polya distribution* when r is a real number. In Fig. 4.2 the negative-binomial probability distribution is depicted for different values of the parameter r, among them $r = 1$ which leads to the geometric distribution, and models the number of failures before the first success. The plot shows that the negative-binomial distribution can take on a variety of shapes, ranging from very skewed to nearly symmetric, depending on the value of r.

This kind of generalized Poisson distribution is frequently used for overdispersed count data. Overdispersion is the condition by which data appear more dispersed than expected under a reference model, typically based on the binomial or Poisson distribution for counting data. If the reference model is Poisson, a number of alternative model formulations are available to increase the dispersion. For example, zero-inflated models add a proportion of zeros to the zeros of a Poisson, while hurdle models are two-part models where zeros and non-zeros are generated by different stochastic processes (see Cameron and Trivedi [4] for details).

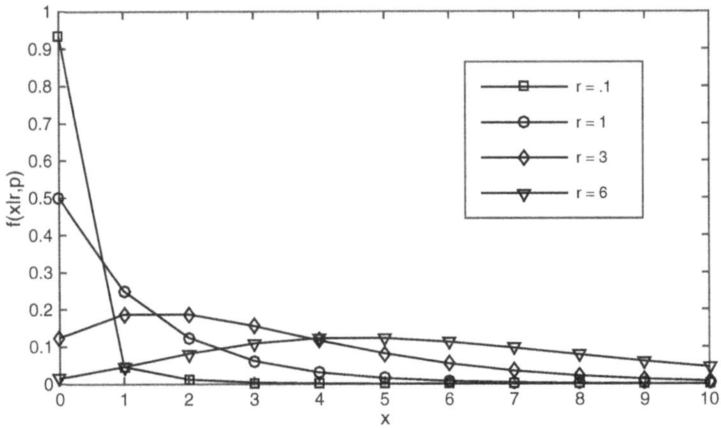

Fig. 4.2 Negative-binomial probability distribution for different values of the parameter r

An alternative approach is to derive from the reference distribution another one that exhibits increased dispersion; for example, by mixing a Poisson process with a gamma distribution representing the distribution of the Poisson parameter, we recover, as pointed out before, the negative-binomial distribution which is overdispersed with respect to the Poisson. On the other hand, Joe and Zhu [13] introduced a generalized Poisson mixed model providing an alternative to the negative-binomial distribution; this model has a scale parameter, and it is heavier in the tail than the negative-binomial one and easily reduces to the Poisson distribution. Consul and Jain [5] proposed a new generalization of the Poisson distribution with two parameters λ_1 and λ_2, obtained as a limiting form of the generalized negative binomial distribution. This model shows a variance of the distribution that is greater than, equal to or smaller than the mean according as λ_2 is positive, zero, or negative and gives a very close fit to binomial, Poisson and negative-binomial data.

4.1.2 Compound Poisson Distribution

A compound Poisson distribution is the probability distribution of the sum of a number of independent identically distributed random variables where the number of terms to be added is itself a Poisson random variable. The resulting distribution can be either a continuous or a discrete distribution.

Definition 4.1.2 (compound Poisson distribution) Given $N \sim Poi(\lambda)$, and $X_i, i = 1, \ldots, N$ i.i.d. random variables and also independent of N, then the probability distribution of the sum of these random variables is a well-defined distribution called compound Poisson distribution:

$$\sum_{i=1}^{N} X_i \sim CPoi(m_{CPoi}, \sigma^2_{CPoi}),$$

where $m_{CPoi} = \mathbb{E}(N)\mathbb{E}(X)$ and $\sigma^2_{CPoi} = \mathbb{E}(N)\mathbb{E}(X^2)$. The probability distribution can be recovered in terms of characteristic functions as $\varphi_Y(t) = e^{\lambda(\varphi_X(t)-1)}$, where $Y = \sum_{i=1}^{N} X_i$.

The discrete compound Poisson distribution is widely used in actuarial science for modelling the distribution of the total claim amount of the sum of a Poisson number N of independent and identically distributed claim amounts $X_i, i = 1, \dots, N$.

Remark 4.1.1 A special case is represented by the *stuttering Poisson distribution* which results from a compound Poisson distribution where $X_i, i = 1, \dots, N$ are nonnegative discrete random variables; this kind of model is used to represent batch arrivals such as in a bulk queue (see Kemp 15). If, for example, X is either an exponential distribution or a gamma one, then the conditional distributions of $Y|N$ are gamma distributions in which the shape parameters are proportional to N. These models are used in Hydrology to model the total rainfall in a day, where each day contains a Poisson-distributed number of events each of which provides an amount of rainfall which has an exponential distribution (see, e.g., Revfeim 21). Thompson [22] applied the same model to monthly total rainfalls.

Example 4.1.1 (Tweedie's compound Poisson model) Tweedie [23] seems to have been the first to study the compound Poisson model with gamma severities in insurance applications. Let i be the accident years and j be the development periods. The number of payments r_{ij} are realizations of random variables R_{ij} that are independent and Poisson-distributed with parameter $\lambda_{ij} w_i > 0$ for a weight $w_i > 0$ that is an appropriate measure for the volume. Appropriate weights can represent the number of policies or the total number of claims incurred. Moreover the size of individual payments $X_{ij}^{(k)}$ are stochastically modelled as independent and gamma distributed variables with mean $\alpha_{ij} > 0$ and shape parameter $\theta > 0$. Assuming that R_{ij} and $X_{ij}^{(k)}$ are independent for all indices, the incremental payments C_{ij} for claims with origin in the accident year i during the development period j is defined as

$$C_{ij} = \mathbf{1}_{R_{ij}>0} \sum_{k=1}^{R_{ij}} X_{ij}^{(k)}.$$

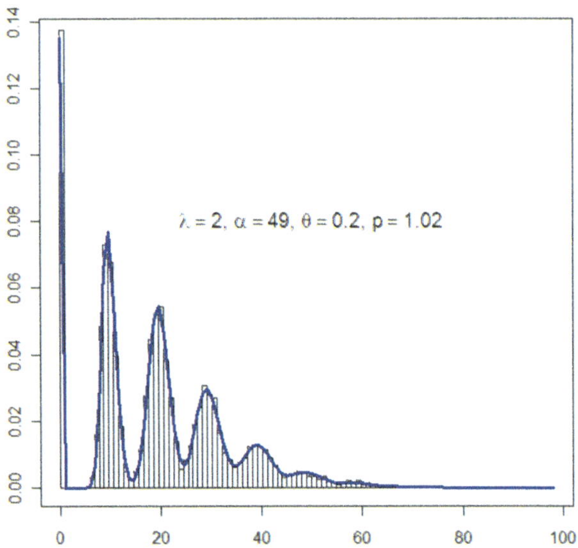

Fig. 4.3 Tweedie probability distribution for given values of the parameters

It is easy to see that conditionally on R_{ij}, the random variable C_{ij} is gamma distributed with mean $R_{ij}\alpha_{ij}$ and shape parameter $R_{ij}\theta$ (see Jørgensen and De Souza 14). In Wüthrich [24] a clever parametrization for Tweedie's compound Poisson model is chosen in such a way that the model belongs to the exponential dispersion family[1] with variance function $V(\mu) = \mu^p$, $p \in (1, 2)$, and dispersion ϕ (see McCullagh and Nelder [18] for detail on the exponential dispersion model). For a constant dispersion ϕ, the model closes the gap between the overdispersed Poisson model (see Renshaw and Verrall 20) and the Gamma model (see Mack 17), corresponding to the extreme cases $p = 1$, implying a linear variance function, and $p = 2$, respectively.

If we assume to have stochastic severities $X \sim \Gamma(\alpha, \theta)$, the dispersion parameter is hence $\phi = \frac{\lambda^{1-p}(\alpha\theta)^{2-p}}{2-p}$, where $p = \frac{\alpha+2}{\alpha+1}$. The resulting Tweedie density is depicted in Fig. 4.3; it shows a multimodal feature which is important to represent particular multifractal phenomena.

[1]This family is a set of probability distributions that represents a generalization of the natural exponential family. A lot of very common probability distributions belong to the class of exponential dispersion family, among them are normal distribution, binomial distribution, Poisson distribution, negative-binomial distribution, gamma distribution, inverse Gaussian distribution, and Tweedie distribution.

4.2 Counting Process: Fundamentals in Literature

We observe that when one talks about counting variable one is usually static in time. On the other hand, if we are interested in the temporal dynamic of a counting random variable, we need to consider a family of such variables defining a counting process.

Definition 4.2.1 (Poisson process) The collection $\{N(t), t > 0\}$ where each $N(t)$ is a Poisson distributed random variable with mean λt that represents the number of event occurrences in $[0, t]$, is a nondecreasing, integer-valued, continuous-time random process called *Poisson process*. This process is characterized by the properties of independent increments for nonoverlapping intervals and of stationary increments, i.e., the distribution of the number of event occurrences is the same in any interval of length t. A Poisson process is homogeneous/inhomogeneous depending on the intensity is constant/time-varying.

In Fig. 4.4 is depicted a simulated trajectory of a Poisson process composed by 500 variates with k randomly varying between 1 and 10 and $\lambda = 12$.

Consider the time T, i.e., the first jump time in a Poisson process where p is the probability of an event occurrence in each subinterval and λ is the rate of event occurrence per subinterval of length $\frac{t}{n}$. We have that

$$\mathbb{P}(T > t) = (1 - p)^n = \left(1 - \frac{\lambda t}{n}\right)^n \rightarrow e^{-\lambda t} \text{ for } n \rightarrow \infty.$$

The random variable T is then an exponential random variable with parameter λ. Since the times between event occurrences in the underlying binomial process are independent geometric random variables, the sequence of inter-event times in a

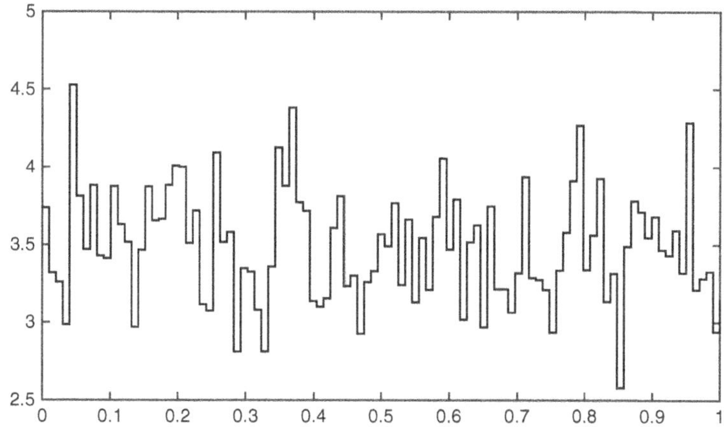

Fig. 4.4 Poisson process simulation, $\lambda = 12$

Poisson process form an i.i.d. sequence of exponential random variables with mean λ^{-1}.

A generalization of an inhomogeneous Poisson process is a Cox process, defined as follows.

Definition 4.2.2 (Cox process) A Cox process is an inhomogeneous Poisson process in which the intensity is allowed to be random with the restriction that conditional on the realization of λ, N is an inhomogeneous Poisson process.

In the same way, a collection of compound Poisson random variables defines a compound Poisson process. We provide the following definition:

Definition 4.2.3 (compound Poisson process) A compound Poisson process with parameter $\lambda > 0$ and jump size distribution G is a continuous-time stochastic process $\{Y(t), t > 0\}$ given by

$$Y(t) = \sum_{i=1}^{N(t)} X_i,$$

where the sum is by convention equal to zero as long as $N(t) = 0$. Here $\{N(t), t > 0\}$ is a Poisson process with parameter λ and $\{X_i, i \geq 1\}$ is a set of independent and identically distributed random variables, with distribution function G, which are also independent of the Poisson process $\{N(t), t > 0\}$.

4.2.1 Counting Processes in Credit Risk Models: The Intensity-Based Approach

The two main types of models that attempt to describe default processes for defaultable financial instruments are usually referred to as structural and reduced-form or intensity models. Structural models explain default events and determine the time of default as a function of the evolution of the firms' structural variables. Intensity models, on the other hand, do not consider the relation between default and firm value but simply define the default event as the first jump of an exogenously given jump process. In other terms, intensity-based approach, in contrast with structural approach, does not formulate economic arguments about firm defaults in term of fundamental firm variables but models default as some unpredictable Poisson-like event.

Letting $T_1, \ldots, T_n$ be the arrival time of default event, $N(t) = \sum_i \mathbf{1}_{T_i \leq t}$ counts the number of default arrivals in time interval $[0, t]$. Depending on the definition of the intensity or hazard rate, we know that the process $N = (N(t))_{t \geq 0}$ with independent increments $N(t) - N(s)$, for $s < t$ is

- A homogeneous Poisson process if λ is constant, i.e.,

$$\mathbb{P}(N(t) - N(s) = k) = \frac{1}{k!}(\lambda(t-s))^k e^{-\lambda(t-s)}$$

- A inhomogeneous Poisson process if $\lambda(t)$ is a determinist function of time, i.e.,

$$\mathbb{P}(N(t) - N(s) = k) = \frac{1}{k!}\left(\int_s^t \lambda(u)du\right)^k e^{-\int_s^t \lambda(u)du}$$

- A Cox process if the intensity is a stochastic process as, for example, it is driven by the following basic affine dynamic:

$$d\lambda_t = a(b - \lambda_t)dt + \sigma\sqrt{\lambda_t}dW_t + dJ_t$$

  where W is a Brownian motion, J is an independent jump process with Poisson arrival intensity c and exponential jump size distribution with mean d, a is the mean-reversion rate, and σ is the diffuse volatility. This process is mean-reverting because the intensity tends to its long-run mean that is given by $b + \frac{cd}{a}$.

In the intensity-based approach, the default time is set equal to the first jump time of the process N. Then $\tau = T_1$ is exponentially distributed with parameter equal to the intensity of process N and the default probability becomes

- Homogeneous Poisson: $\mathbb{P}(\tau \leq t) = 1 - e^{-\lambda t}$
- Inhomogeneous Poisson: $\mathbb{P}(\tau \leq t) = 1 - e^{-\int_0^t \lambda(u)du}$
- Cox process: $\mathbb{P}(\tau \leq t) = 1 - \mathbb{E}\left(e^{-\int_0^t \lambda(u)du}\right)$

To introduce default correlations, we refer to the three main approaches to incorporate credit risk correlations among firms within the framework of reduced models. The first one, known as conditionally independent defaults models, assumes that default intensities are dependent on a set of common factors and on a firm specific factor such that the default intensities are independent, conditionally to the common state variables. The default intensity of firm i can be specified as

$$\lambda_{i,t} = a_{0,t} + a_{1,t}X_{1,t} + \ldots + a_{J,t}X_{J,t} + \varepsilon_{i,t},$$

where $a_{j,t}$, $j = 1, \ldots, J$ are some deterministic coefficients explaining the dependence on common factors $X_{j,t}$ while $\varepsilon_{i,t}$ is the idiosyncratic factor independent across firms. We refer to Duffee [8], Zang [25], Driessen [7], and Elizalde [10] as examples of the formalization and the estimation of this class of model. The main drawback of these models concerns the comparison between empirical default correlations and the theoretical ones that seems too low. Duffie and Singleton [9] proposed a model to correct this problem by introducing either joint jumps in the firms' default processes or common default events.

The second way to introduce default dependence between firms is an extension of the previous approach to account for the empirical observation of default clustering. The literature of default contagion includes the infectious defaults model (see Davis and Lo 6) and the propensity model (see Jarrow and Yu 12). In the dynamic version of the infectious defaults model, default's occurrences cause an increasing default intensity of all remaining firms; it increases by an enhancement factor $a > 1$ for a period of time that is exponentially distributed and after which the enhancement factor disappears. During the period of augmented intensity, the risk of default contagion is reflected in a higher level of default probabilities of all firms. On the other hand, propensity models account for counterparts risk to explain default clustering. They distinguish between primary and secondary firms; primary ones do not depend on the default of other firms, but their defaults increase the default intensity of secondary firms. Then default intensities of secondary firms are the sum of an independent firm's intensity and several dependent ones representing the previously explained asymmetric dependence. Therefore these kinds of models do not solve the drawbacks of low correlation between primary firms.

The last type of model is represented by copula approach which allows to separate marginal default probabilities from the credit risk dependence structure. Obviously as we can see in the following, the intensity approach is not the only framework where we can use copula function to model the default dependence structure between firms. The joint default probability of default times is represented as

$$\mathbb{P}(\tau_1 \leq T_1, \ldots, \tau_I \leq T_I) = C(F_1(T_1), \ldots, F_I(T_I)),$$

where $\tau_i, i = 1, .., I$ is the default time of i-th firm. The advantage in specifying the dependence structure in terms of copula function relies on the possibility to separate the modelling of marginals from that of the dependence structure represented by the copula function. Here the choice of a copula showing lower tail dependence, as, for example, the t-copula among the elliptical ones and the Clayton copula among the Archimedean ones, allows to represent default clustering periods. Analogously the choice of a proper kind of copula function lets the contagion effects arise endogenously through the dependence structure. Parameter risks arising from the difficulties linked to the rigorous estimation of copula's parameters together with the model risk, i.e., the risks linked to the choice of the proper copula function, are among the principal drawbacks of the copula approach in the modelling of dependent defaults.

4.3 A New Combinatoric Approach for Counting

The aim of this section is to present our original approach to a counting problem. As already reviewed in Part I, our approach is made up by three main ingredients, i.e., the clustering, the copula, and the combinatorics.

Given a set of variables, we proceed following the listed steps:

- The clustering phase: we move forward by grouping the variables in classes through a similarity criterion, and we select/calibrate a dependence structure representing the relations/contagion among classes.
- The scenarios' generation phase: actually we are able to explain all possible scenarios (or the set of most probable ones) related to the occurrences of a random event affecting those classes. As explained in Chap. 3, they are represented by the combinatorial distribution of the event on the set of groups (or by the distorted set of them if we rely our interest to most probable scenarios). Anyway the combinatorial distribution of the random event generates a matrix whose rows identify different scenarios.
- The counting variable setting: in order to count the occurrences of the random event, we need to associate to every row an integer number; this integer is a representation of the counting variable whose probability should be related to the confidence that the corresponding scenario takes place. This probability, as will be explained in more depth later, is evaluated as a volume of the copula function representing the dependence structure of the whole set of classes.

As we can observe, the clustering phase allows for a reduction of the complexity of the problem and lets us manage high-dimensional exercises as well. Obviously the clustering is effective if and only if we are able to calibrate the dependence structure of the random variables without losing important information; for this reason the more homogeneous are the variables inside every group, the more effective is the clustering approach. The dependence structure of the groups is also an important input of the probabilistic description of the counting variable: as a matter of fact, the probability distribution of such a variable is evaluated as a set of volumes of the copula function explaining the groups' connections.

4.3.1 A Counting Variable Linked to a Clusterized Homogeneous Dependence Structure

As an example, let us count a random event attaching a set of random variables whose dependence structure is represented by a *Clusterized Homogeneous Copula* (CHC for short), which has been already defined and discussed in Sect. 2.4.1. This kind of copula is a grouping copula which might be not hierarchical as well. Indeed a CHC, with just one level of dependence (previously noted as 1-CHC), is not hierarchical because the connections among groups are equally strong. Therefore a 1-CHC could be a useful tool to reduce the complexity of the problem.

In the following we assume to work within the homogeneous class of CHC, but this assumption is not restrictive: the same arguments hold true for all kind of clusterized copulas having groups with cardinality ≥ 1. Obviously, if the cardinality is one, this means that we are working with the singular variables instead of grouping them. In this case we no longer have a clusterized copula but a standard operator.

Let us first present the heuristics of our ideas for the extraction of the pdf of a counting variable from the dependence structure of the variables themselves. As showed in what follows, the volume of a copula is involved into the computation of the probability distribution of the counting variable.

We focus on a two-dimensional problem, as, for example, we assume to have a portfolio composed by two loans grouped into two single clusters of equivalent cardinalities $\begin{bmatrix} 1 & 1 \end{bmatrix}$ and having the default probabilities $\begin{bmatrix} v_1 & v_2 \end{bmatrix}$, corresponding to the entries of the vector. We observe that if the dependence structure of the loans is represented by a copula C, the joint probability of defaults, i.e., the probability to register two defaults or that the counting variable r (which counts the number of default) be equal to 2, is

$$P_r(2) = C(v_1, v_2) = V_C(\mathbf{S}),$$

where V_C stands for the volume of copula C in the box $\mathbf{S} = \left[\begin{pmatrix} 0 \\ 0 \end{pmatrix}, \begin{pmatrix} v_1 \\ v_2 \end{pmatrix} \right]$ that can be represented as in Fig. 4.5a and matches the concept of joint probability of both the default events.

In the same spirit, the joint probability of survival, i.e., the probability to register zero defaults, corresponds to the survival copula function:

$$P_r(0) = \overline{C}(v_1, v_2) = V_C(\overline{\mathbf{S}}),$$

where $\overline{C}$ stands for the survival copula and the box $\overline{\mathbf{S}} = \left[\begin{pmatrix} v_1 \\ v_2 \end{pmatrix}, \begin{pmatrix} 1 \\ 1 \end{pmatrix} \right]$. See Fig. 4.5b for a useful representation of such volume.

Now the question is *How do we compute the probability to have only one default?* The expected answer is by a copula volume too, we are now going to detail.

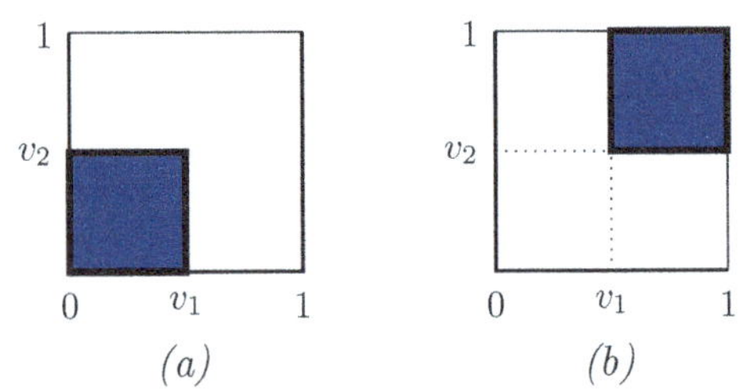

Fig. 4.5 Graphical representation of the volume of copula C in the boxes $\mathbf{S} = [\mathbf{0}, \mathbf{v}]$ and $\mathbf{S} = [\mathbf{v}, \mathbf{1}]$

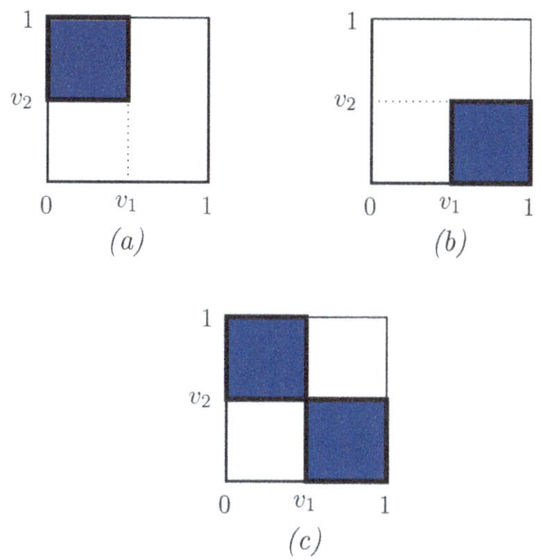

Fig. 4.6 Graphical representation of the volume of copula C in boxes accounting for one default

The probability that the first loan defaults and the second survives is also a volume computed in a particular box:

$$P_r(1,0) = V_C\left[\begin{pmatrix} 0 \\ v_2 \end{pmatrix}, \begin{pmatrix} v_1 \\ 1 \end{pmatrix}\right],$$

that corresponds to the gray area in Fig. 4.6a. In the same way, the probability that the first loan survives and the second defaults is

$$P_r(0,1) = V_C\left[\begin{pmatrix} v_1 \\ 0 \end{pmatrix}, \begin{pmatrix} 1 \\ v_2 \end{pmatrix}\right]$$

that corresponds to the area plotted in Fig. 4.6b.

Then the probability to have one default, no matter which one, is $P_r(1) = P_r(1,0) + P_r(0,1)$, i.e., the total area in Fig. 4.6c.

The probability distribution of the counting variable is then the collection $\{P_r(0), P_r(1), P_r(2))\}$.

The link between the volume of a copula function and the probability distribution of the counting variable is now clear. To specialize our technique for a clusterized copula, i.e., for a CHC, we must generalize the explained procedure accounting for groups instead of single variables. Therefore we first consider the combinatorial distributions of all possible number of events into the groups. We propose an example to explicitly explain the generalized version of probabilities' assessment of a counting variable which involves groups of random variables.

Example 4.3.1 (Counting on a CHC) Consider a portfolio composed by three loans grouped into two clusters with equivalent cardinalities $\begin{bmatrix} 2 & 1 \end{bmatrix}$. The concept of equivalent cardinality allows to work with homogeneous groups thanks to a possible

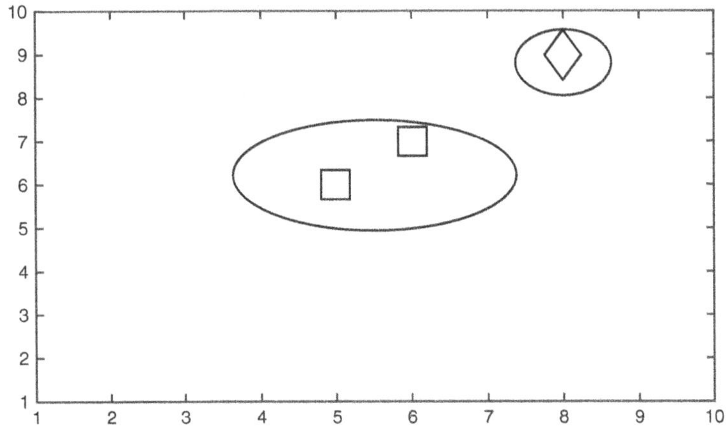

Fig. 4.7 The clustering structure of the loans

compensation of groups' dimension. In Fig. 4.7 is visualized the clustering structure of the loans.

In order to define the scenarios, we need to recover the combinatoric distribution of the default event in the two groups. We focus on a particular event, i.e., the event causing just one default, whose possible corresponding scenarios are represented by the rows of the following matrix:

$$M = \begin{bmatrix} 1\,0\,0 \\ 0\,1\,0 \\ 0\,0\,1 \end{bmatrix}$$

where the first two columns identify the loans placed in the first group, while the third column represents the loan belonging to the second group.

We observe that only the first and the third row of the matrix are relevant since we assume that the groups are homogeneous implying that we must not consider the order in the distribution of the random event into the groups. This way the first two scenarios can be assumed to have the same probability to take place.

Following the same logic explained before, every row is a scenario whose probability to arise corresponds to the volume of a copula function. This volume is evaluated in a box, whose coordinates are defined by the scenario itself. More precisely the probability attributed to the first and the second scenario, i.e., the first and the second row of matrix M, matches the volume:

$$P_r(1,0) = V_{CHC}\left[\begin{pmatrix} 0 \\ v_2 \end{pmatrix}, \begin{pmatrix} v_1 \\ 1 \end{pmatrix}\right],$$

while the probability attributed to the third row now corresponds to

$$P_r(0,1) = V_{CHC}\left[\begin{pmatrix} v_1 \\ 0 \end{pmatrix}, \begin{pmatrix} 1 \\ v_2 \end{pmatrix}\right].$$

Therefore the probability to have one default is given by

$$P_r(1) = 2P_r(1,0) + P_r(0,1),$$

since we can distribute the integer in two ways into the first group (corresponding to the first and the second row of the matrix M), associated to the same probability due to the homogeneous assumption.

We observe that the generalization to the clusterized copulas is trivial because the information we have to take into account refers to the number of defaults recorded in every group; hence the complexity of dimension 2 of the first group is really reduced to 1 as in case of standard copulas.

4.3.2 Clusterized Homogeneous Copulas: CHC-Volume

Here we deal with the class of clusterized copulas whose groups are homogeneous and aggregated taking into account a hierarchy. This class has been defined and discussed in Bernardi and Romagnoli [1]. The problem is simplified with respect to the aggregation discussed in the Part I of the book because we assume the groups are homogeneous.

Proposition 4.3.1 (Clusterized Homogeneous Copula volume) *Given a clusterized Hierarchical copula function CHC with k homogeneous groups of dimension* $m_s, s = 1, .., k$ *respectively such that* $\sum_{s=1}^{k} m_s = n$*, the volume of the k-dimensional box* $\mathbf{S} = [\mathbf{u}, \mathbf{v}]$ *with* $\mathbf{u}, \mathbf{v} \in [0,1]^k$, $\mathbf{u} \leq \mathbf{v}$, *may be represented as*

$$V_{CHC}(\mathbf{S}) = \sum_{i=0}^{n} (-1)^i \sum_{j=1}^{\hat{D}^c(i,k)} CHC(\mathbf{c}(\mathbf{p}_{i,k}(j)))$$

where $\hat{D}^c(i,k)$ *denotes the number of compatible combinatorial distribution (c.c.ds in the following)*[2] *of the integer i into k groups,* $\mathbf{p}_{i,k}(j)$ *is a* $k \times \max(m_s, s = 1, \ldots, k)$ *matrix of zeros and ones corresponding to the j-th c.c.d. of i ones into*

[2]We refer to $\hat{D}(i,k)$ as the number of the ways in which one can distribute the integer i into k groups without taking into account the order of the groups (c.ds for short). If we take into account the groups' cardinalities, then we refer to $\hat{D}^c(i,k)$ as the number of the compatible (for cardinalities' groups) combinatorial distributions.

k groups of dimensions $m_s, s = 1, \ldots, k$, *and* $\mathbf{c}(\mathbf{p}_{i,k}(j))$ *is a* $k \times \max(m_s, s = 1, \ldots, k)$ *matrix such that*

$$c_{s,w} = v_s \quad \textit{if} \quad p_{s,w} = 0$$

and

$$c_{s,w} = u_s \quad \textit{if} \quad p_{s,w} = 1,$$

where u_s, v_s *are the* s*-th element of the corresponding vector and* $c_{s,w}$ *denote the* (s, w)*-th element of the corresponding matrix and* $CHC(\mathbf{c}(\mathbf{p}_i(j)))$ *is the clusterized homogeneous copula computed for the* j*-th c.c.d. accounting for the permutation of the elements into the groups.*

Remark 4.3.1 We can observe that $\hat{D}(i,k)$ can be computed by a number of permutations with repetitions, i.e.,

$$\hat{D}(i,k) = \binom{n+k-1}{k-1}$$

For example, if we assume two groups each one with cardinality three, the number of c.ds of $i = 1$ is $\hat{D}(1,2) = \hat{D}^c(1,2) = 2$ since all the c.ds are compatibles with the groups' cardinalities. In this framework these two c.c.ds correspond to the following situations:

1. the integer 1 is placed in the first group, independently of the position (for convention in the first place);
2. the integer 1 is placed in the second group, independently of the position (for convention in the first place).

We observe that the order within the groups doesn't matter because they are homogeneous.

If we order these c.c.ds, then the c.c.ds for $i = 1$ distributed into two groups each one with cardinality three corresponds to the following matrices:

$$\mathbf{p}_{1,2}(1) = \begin{bmatrix} 1\,0\,0 \\ 0\,0\,0 \end{bmatrix} \qquad \mathbf{p}_{1,2}(2) = \begin{bmatrix} 0\,0\,0 \\ 1\,0\,0 \end{bmatrix}$$

where the first rows depict the distribution inside the first group, while the second row stands for the distribution inside the second class.

Therefore the computation of the CHC-volume in the box $\mathbf{S} = \left[\begin{pmatrix} u_1 \\ u_2 \end{pmatrix}, \begin{pmatrix} v_1 \\ v_2 \end{pmatrix}\right]$ proceeds with the following steps:

1. Evaluate every scenario and represent it as an ordered matrix $\mathbf{p}_{i,k}(j), i = 0, \ldots, 6, k = 2$, where j enumerates the scenario following a column's order.
2. Apply the operator c to every matrix $\mathbf{p}_{i,k}(j), i = 0, \ldots, 6, k = 2, \forall j$.

3. Compute the CHC having marginals equal to the entries of every matrix $\mathbf{c}(\mathbf{p}_{i,k}(j)), i = 0, \ldots, 6, k = 2, \forall j$.

The operator c acts substituting the entries of the matrix $\mathbf{p}_{i,k}(j), i = 0, \ldots, 6, k = 2, \forall j$ as described in Proposition 4.3.1. Hence the transformations of the scenarios corresponding to $i = 1$ distributed into two groups each one with cardinality three leads to the following matrices:

$$\mathbf{c}(\mathbf{p}_{1,2}(1)) = \begin{bmatrix} u_1 & v_1 & v_1 \\ v_2 & v_2 & v_2 \end{bmatrix} \qquad \mathbf{c}(\mathbf{p}_{1,2}(2)) = \begin{bmatrix} v_1 & v_1 & v_1 \\ u_2 & v_2 & v_2 \end{bmatrix}$$

Finally we need to compute the copulas:

$$CHC\left(\mathbf{c}(\mathbf{p}_{1,2}(1))\right) = 3CHC(u_1, v_1, v_1, v_2, v_2, v_2)$$
$$CHC\left(\mathbf{c}(\mathbf{p}_{1,2}(2))\right) = 3CHC(v_1, v_1, v_1, u_2, v_2, v_2)$$

which account for the permutations of the elements into the groups.

As already observed in Part I of the book, a special CHC-volume is represented by the Survival copula. As a matter of fact the computation of the survival copula function is a particular case of the volume computation.

Corollary 4.3.2 (Survival copula function in n dimension) *In the same setting of Proposition* 4.3.1, *the survival of a CHC with marginals* $\mathbf{u} \in [0, 1]^k$ *may be represented as*

$$V_{CHC}(\mathbf{S}) = \sum_{i=0}^{n}(-1)^i \sum_{j=1}^{\hat{D}^c(i,k)} CHC(\mathbf{c}(\mathbf{p}_{i,k}(j)))$$

where the box $\mathbf{S} = [\mathbf{u}, \mathbf{e}]$ *and where* $\mathbf{e} = (1, 1, \ldots, 1)$ *is the unit vector in* $\mathbb{R}^k$.

4.3.3 Preparing the CHC-Computation

Giving the c.c.ds' matrix, we compute the copula function into the groups and so we create a matrix of dimension equal to the number of groups $\times$ the maximum dimension of the groups (this is the matrix $\mathbf{p}_{i,k}(j)$ in Proposition 4.3.1) where we distribute i ones into k groups according to the j-th c.c.d.. For example, if the second row of the c.c.ds' matrix for the integer 4 into 3 groups with the same dimension equal to 3 is the following:

$$[0\ 1\ 3]$$

and we order following the row's order,[3] the corresponding 0-1 matrix is

$$\mathbf{p}_{4,3}(2) = \begin{bmatrix} 0\,0\,0 \\ 1\,0\,0 \\ 1\,1\,1 \end{bmatrix}$$

Then we must substitute the elements of the matrix $\mathbf{p}_{4,3}(2)$ with the coordinates requested to compute our object, following the method suggested by Proposition 4.3.1. If we want to compute the copula's volume, we must substitute the element of $\mathbf{p}_{4,3}(2)$ with the corresponding element of the vectors defining the box. For example, if the box is $S = [col(0.3 \times \mathbf{I}), col(0.5 \times \mathbf{I})]$ where $\mathbf{I}$ is the identity matrix in $\mathbb{R}^3$, the matrix $\mathbf{p}_{4,3}(2)$ becomes

$$\mathbf{c}(\mathbf{p}_{4,3}(2)) = \begin{bmatrix} 0.5\;0.5\;0.5 \\ 0.3\;0.5\;0.5 \\ 0.3\;0.3\;0.3 \end{bmatrix}$$

The following MatLab™ code explains the procedure.

```
% This function returns a matrix A (it's c(p_ik(j))) from the
% 0-1 matrix np according to the following rule: if we hit on
% a zero, we substitute the corresponding element of vector v,
% whereas if we hit on a one we substitute the corresponding
% element of vector u.

% np is a 0-1 matrix (each row is a c.c.d.)
% sizecluster is the vector of the groups's cardinalities
% u and v are the vectors of volume's coordinates

function[Amat]=mmv(np,u,v,sizecluster)
n = size(v,2);
for i=1:size(np,2)
A(i,:)= [ones(1,np(1,i)) ones(1,max(sizecluster)-np(1,i))*0];
end
for i=1:size(A,1)
for j=1:size(A,2)
if (A(i,j)==0)
        A(i,j)= v(1,i);
else
        A(i,j)= u(1,i);
end
end
end
Amat=A;
```

[3]Following this calling order, the c.c.d. considered here is the second one, i.e. $j = 2$.

4.3.4 CHC and CHY Computation

We compute the copula function into the groups; we aggregate the marginal probabilities with an Archimedean copula (in the following MatLab™ code we compute the Clayton's copula function with parameter equal to the EQDP of the corresponding group). This must be done for each row of the c.c.d. matrix.

```
% This function computes the Clayton's hierarchical second level
% copula, and the product of binomial coefficients allows to
% take account for the position into the groups.

% alpha is the vector of the parameters inside the groups
% alpha_0 is the parameters among groups
% numones is the c.c.ds' matrix for the same number of
% occurences into the groups
% sizecluster is the vector of the groups' cardinalities
% u and v are the vectors of volume's coordinates

function[HCM,pfactor]= claytonLevel2(numones,u,v,sizecluster...
    ,alpha,alpha_0)
numclusters=size(numones,2);

% Given a row corresponding to a c.c.d., this function creates
% a matrix with a number of rows equal to the number of groups
% and a  number of columns equal to the max cardinality of
% groups (it's p_ik(j))

Amat=mmv(numones,u,v,sizecluster);
for j=1:numclusters
for i=1:sizecluster(1,j)

% The function G_Clayton computes the copula's generators
% for each element of matrix Amat and for the parameter of
% the corresponding group.

Cmat(j,i)=G_Clayton(Amat(j,i),alpha(1,j));
end
end
Cmat=[Cmat(:,:)];

% Here we sum over the rows of matrix Cmat till the cardinality
% of the corresponding group.

for j=1:numclusters
a(j)=sum(Cmat(j,1:sizecluster(1,j)));
sumG(j,1) =a(j);

% The function Comp_G_Clayton computes the composed function
% \diamondsuit for the elements of vector a.

arg(j,1) =Comp_G_Clayton(sumG(j,1),alpha_0,alpha(1,j));
```

```
end
arg=[arg(:,:)];

% The function Inv_G_Clayton  computes the inverse of the
% generator function for the sum of the elements of vector arg.

sumCompG=sum(arg);
HCM=Inv_G_Clayton(sumCompG,alpha_0);

% We compute the product of the binomial coefficient for
% each row, i.e. for each c.c.d.

for i=1:numclusters
factorS(i,1)= nchoosek(sizecluster(1,i),numones(1,i));
end
factorS=[factorS(:,1)];
pfactor=prod(factorS);
```

Finally we compute the copula function at the first level, i.e., we aggregate the second-level copula functions with an Archimedean copula with parameter equal to the corresponding EQDP, accounting for the position of the element into the groups.

4.3.5 The Volume of a Clusterized Copula: CHC and CHY

Copula volume is the main tool of the new algorithm. Here we detail the MatLab™ code to compute it.

```


% alpha_0 stands for the Archimedean parameter \textit{between}.
% alpha is the vector of Archimedean parameters \textit{within}.
% num is the vector of groups' cardinalities

function [VOLUME]= VOL(un,vn,c,alpha,alpha_0,num)
dim=size(un,2);

% this is a row vector of integers from 0 to 2^(dim)-1
Aloc=0:((2^dim) -1);
% transform a row into a column
Aloc=Aloc';
% write the integers in  binary representation
q=quantizer([dim+1 0]);
Aloc=num2bin(q,Aloc);
% transform the vector into a matrix (0=48 e 1=49)
Aloc = double(Aloc);
% erase the first column
Aloc(:,1)=[];
% 48 and 49 become the corresponding entries of vn and un,
% respectively
B=matrice(Aloc,un,vn);
% the case c selects the kind of copula (c=1
```

```
% Archimedean copula; c=2 Archimedean hierarchical).
% here we generate a vector with sign depending on the number
% of ones
            switch c
    case{1}
        A=copula_vect(B,alpha_0,Aloc);
    case{2}
        A = H_mat(B,alpha,alpha_0,num,Aloc);
end
% sum the entries of the vector
VOLUME=volume(A);
```

4.3.6 Pdf of a Counting Variable Linked to a CHC: A Formal Approach

In the following we explain how to recover the pdf of a counting variable referred to a random event which hits a set of variables. We assume that a *clusterized homogeneous copula* represents the dependence structure of this set of variables. Given that our procedure to extract the pdf of a counting variable involves the volume of a CHC in n dimensions, we refer to the previous representation of this volume and introduce the following definition.

Definition 4.3.1 (Counting variable linked to a n-dimensional CHC) The counting variable linked to the $CHC(\mathbf{c}(\mathbf{p}_{l,k}(j)))$ is given by

$$r(j) = \sharp(\mathbf{p}_{l,k}(j)), \forall j$$

which counts the number of elements equal to one in the matrix $\mathbf{p}_{l,k}(j)$, $\forall j$, where j stands for the j-th combinatoric compatible[4] distributions (c.c.d. for short). Clearly we have $r(j) = l, \forall j$.

Our aim is to recover the pdf of the counting variable linked to the n-dimensional CHC. To achieve this we observe that a number l corresponds, through his probability function, to a number in $[0, 1]$ by considering all the c.c.ds of l ones into k groups (we enumerate this c.c.d. with the index j).

The j-th c.c.d., corresponding to the matrix $\mathbf{p}_{l,k}(j)$, generates a pair of coordinates,[5] i.e., the j-th c.c.d. generates the box $\mathbf{S}^j$ where we compute the volume. Therefore, given l and k, the algorithm calls for

1. Determining the scenarios in term of matrices $\mathbf{p}_{l,k}(j), \forall j$
2. Corresponding to every matrix of order j, $\mathbf{p}_{l,k}(j)$, a box $\mathbf{S}^j, \forall j$

[4]The concept of compatibility allows to restrict our interest to the scenarios distributing a number of events that doesn't exceed the cardinality of the group itself.

[5]The rule which explains how the coordinates are generated is presented in the following Corollary.

3. Evaluating the volume of the CHC in the box $\mathbf{S}^j, \forall j$

The sum of the volumes computed for all the coordinates generated by all the c.c.ds for the same number of distributed ones, i.e., for the same l ones into k groups, represents the probability to count l ones, or better to have l events whose marginal probabilities are the centroids of the groups.

In Bernardi and Romagnoli [2] is provided an algorithm to extract from a CHC the probability information about the counting variable. This algorithm is based on the concept of volume of a CHC (see Proposition 4.3.1 for a formal definition).

Corollary 4.3.3 (pdf of the counting variable linked to a n-dimensional CHC) *The pdf of the counting variable linked to the n-dimensional CHC, with k homogeneous groups of equivalent dimension $m_s, s = 1, .., k$, respectively, such that $\sum_{s=1}^{k} m_s = n$ and with centroids $\mathbf{u} \in [0, 1]^k$ is the function $P_r : [0, n] \to [0, 1]$ such that*

$$P_r(l) = \sum_{j=1}^{\hat{D}^c(l,k)} V_{CHC}(\mathbf{S}^j(\mathbf{p}_{l,k}(j)))$$

where $\mathbf{p}_{l,k}(j)$ is a $k \times \max(m_s, s = 1, \dots, k)$ matrix of zeros and ones corresponding to the j-th c.c.d. of l ones into k groups of dimensions $m_s, s = 1, \dots, k$ and finally accounting for the permutation of the elements into the groups and where $V_{CHC}(\mathbf{S}^j)$ is the volume of the CHC computed for the box $\mathbf{S}^j = [\mathbf{u}^j, \mathbf{v}^j] \in \mathbb{R}^z \times \mathbb{R}^z$ with $\max(z) = 2k$, determined for the j-th c.c.d. of l ones into k groups following the rûle:

- *If $p_{s,w} = 0$, $u^j_{s,w} = u_s$ and $v^j_{s,w} = 1$ for $s = 1, \dots, k$ and $w = 1, \dots, m_s$ while if $p_{s,w} = 1$, $u^j_{s,w} = 0$ and $v^j_{s,w} = u_s$ for $s = 1, \dots, k$ and $w = 1, \dots, m_s$.*
- *Since the values of the coordinates corresponding to each $p_{s,w}$ may be only two, we correspond to each coordinate a two-dimensional vector such that*

$$u^j_{s,w}, \forall w \to \mathbf{u}^j_s = [u^j_{s,1}, u^j_{s,2}] \quad v^j_{s,w}, \forall w \to \mathbf{v}^j_s = [v^j_{s,1}, v^j_{s,2}],$$

where $(u^j_{s,1}, v^j_{s,1})$ represents the coordinates of the not empty 0-elements set[6] and $(u^j_{s,2}, v^j_{s,2})$ the ones of the not empty 1-elements set in the matrix $\mathbf{p}_{l,k}(j)$.
- *Finally we define the box through the coordinates' vectors $[\mathbf{u}^j_s, \mathbf{v}^j_s]$ as*

$$\mathbf{u}^j = row(\mathbf{u}^j_s) \quad \mathbf{v}^j = row(\mathbf{v}^j_s),$$

[6] In the case of empty set, we delete the correspondent vector element. For example, if the set of elements equal to one into the s-th group and for the j-th c.c.d. is empty, we correspond only $u^j_{s,2}$ to $u^j_{s,w}, \forall w$ and $v^j_{s,2}$ to $v^j_{s,w}, \forall w$.

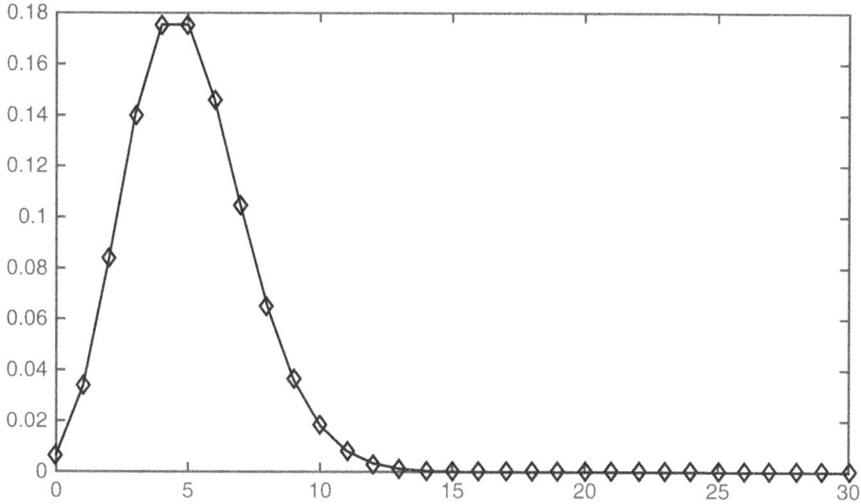

Fig. 4.8 Pdf of a counting variable with range from 0 to 30 linked to a 1-CHC structure

where the function row : $\mathbb{R}^k \times \mathbb{R}^m \to \mathbb{R}^{k\times m}$ *is defined as*

$$row\begin{pmatrix} x_1^1 & \cdots & x_1^m \\ \vdots & \vdots & \vdots \\ x_k^1 & \cdots & x_k^m \end{pmatrix} = \begin{bmatrix} x_1^1 & \cdots & x_1^m & \cdots & x_k^1 & \cdots & x_k^m \end{bmatrix}.$$

In Fig. 4.8 is represented the pdf of a counting variable defined on a range from 0 to 30 and recorded over a 5 groups of dimension 6 whose dependence structure is a 1-CHC.

Referring to the evaluation of the pdf of the counting variable explained in Corollary 4.3.3, in the following example we lay out the computational phase of the coordinates defining the box.

Example 4.3.2 (Box's definition) To compute the pdf of the counting variable linked to a CHC, we must firstly compute the c.c.ds of an integer into the groups already determined by the clustering procedure. We must consider the c.c.ds for all the integers from zero to the dimension of the problem. Each c.c.d.[7] corresponds to a box, i.e., a pair of coordinates, where we must compute the volume as explained in Corollary 4.3.3.

[7]We assume that this c.c.ds are compatibles with the equivalent cardinalities of groups determined by the clustering method and then compensated for the homogeneous assumption.

For example, if we consider the distribution of the integer 2 into three groups with the same dimension equal to 2, we identify the following scenarios:

$$\begin{bmatrix} 1\,1\,0 \\ 0\,1\,1 \\ 1\,0\,1 \end{bmatrix},$$

whose rows generate different boxes $\mathbf{S}^j$, $j = 1, 2, 3$, where j traces the order of the row in the matrix of the combinatoric distributions. As a matter of fact, the first row is

$$[1\;1\;0] \rightarrow \mathbf{p}_{2,3}(1) = \begin{bmatrix} 1\,0 \\ 1\,0 \\ 0\,0 \end{bmatrix},$$

whose coordinates, defining the box $\mathbf{S}^1$ are

$$\mathbf{u}_2^1 = \begin{bmatrix} 0 & v_1 \\ 0 & v_2 \\ v_3 & - \end{bmatrix} \qquad \mathbf{v}_1^2 = \begin{bmatrix} v_1 & 1 \\ v_2 & 1 \\ 1 & - \end{bmatrix},$$

or better

$$\mathbf{u}^1 = \begin{bmatrix} 0 \; v_1 \; 0 \; v_2 \; v_3 \end{bmatrix}$$
$$\mathbf{v}^1 = \begin{bmatrix} v_1 \; 1 \; v_2 \; 1 \; 1 \end{bmatrix},$$

where $\begin{bmatrix} v_1 \; v_2 \; v_3 \end{bmatrix}$ are the event's probabilities of the centroids. We omitted the second component of the third group's coordinates because it is actually homogeneous in this scenario. It's important to observe that the box defines a new cardinality structure since the first and the second group can't be considered homogeneous.

Similarly the second row of the c.c.ds' matrix is the following:

$$[0\;1\;1] \rightarrow \mathbf{p}_{2,3}(2) = \begin{bmatrix} 0\,0 \\ 1\,0 \\ 1\,0 \end{bmatrix},$$

and the corresponding coordinates defining the box $\mathbf{S}^2$ are

$$\mathbf{u}_2^2 = \begin{bmatrix} v_1 & - \\ 0 & v_2 \\ 0 & v_3 \end{bmatrix} \qquad \mathbf{v}_2^2 = \begin{bmatrix} 1 & - \\ v_2 & 1 \\ v_3 & 1 \end{bmatrix},$$

where we omitted the second component of the first group's coordinates because it is actually homogeneous in this scenario. The coordinates of the box can be written in a row representation, as follows:

$$\mathbf{u}^2 = \begin{bmatrix} v_1 \; 0 \; v_2 \; 0 \; v_3 \end{bmatrix}$$
$$\mathbf{v}^2 = \begin{bmatrix} 1 \; v_2 \; 1 \; v_3 \; 1 \end{bmatrix},$$

where $\begin{bmatrix} v_1 \; v_2 \; v_3 \end{bmatrix}$ are the event's probabilities of the centroids. It's important to observe that the box defines a new cardinality structure since the second and the third group can't be considered homogeneous.

Finally the third row corresponds to the following scenario:

$$[1\;0\;1] \rightarrow \mathbf{p}_{2,3}(3) = \begin{bmatrix} 1\;0 \\ 0\;0 \\ 1\;0 \end{bmatrix},$$

whose coordinates, defining the box $\mathbf{S}^3$ are

$$\mathbf{u}_2^3 = \begin{bmatrix} 0 & v_1 \\ v_2 & - \\ 0 & v_3 \end{bmatrix} \qquad \mathbf{v}_3^2 = \begin{bmatrix} v_1 & 1 \\ 1 & - \\ v_3 & 1 \end{bmatrix},$$

or better

$$\mathbf{u}^3 = \begin{bmatrix} 0 \; v_1 \; v_2 \; 0 \; v_3 \end{bmatrix}$$
$$\mathbf{v}^3 = \begin{bmatrix} v_1 \; 1 \; 1 \; v_3 \; 1 \end{bmatrix},$$

where $\begin{bmatrix} v_1 \; v_2 \; v_3 \end{bmatrix}$ are the event's probabilities of the groups' centroids. It's important to observe that the box defines a new cardinality structure since the first and the third group can't be considered homogeneous.

Moreover we observe that the volume of the box $\mathbf{S}^j = [\mathbf{u}^j, \mathbf{v}^j]$, $j = 1, 2, 3$ must be computed for a clusterization into 5 groups whose cardinalities is, for example, $\begin{bmatrix} 2 \; 1 \; 1 \; 1 \; 1 \end{bmatrix}$ if $j = 2$. Each c.c.d. modifies the structure of the problem in term of the number and the cardinality of groups.

Finally we observe that the sum of the volumes of the boxes for all $j = 1, 2, 3$ represents the probability that the counting variable is equal to 2, i.e.,

$$P_r(2) = V_{CHC}(\mathbf{S}^1) + V_{CHC}(\mathbf{S}^2) + V_{CHC}(\mathbf{S}^3).$$

Therefore the set of $P_r(j)$, $j = 0, 1, ..6$ defines the pdf of the discussed counting variable.

The pdf of the counting variable already discussed is useful to recover the corresponding cdf, as explained in the following.

Corollary 4.3.4 (cdf of the counting variable linked to the n-dimensional CHC) *In the same setting of Corollary* 4.3.3, *the cdf of the counting variable linked to the n-dimensional CHC, with k homogeneous groups of equivalent dimensions m_s, $s = 1, .., k$, respectively, such that $\sum_{s=1}^{k} m_s = n$ and with centroids $\mathbf{u} \in [0, 1]^k$, is the function $F_r : [0, n] \rightarrow [0, 1]$ such that*

$$F_r(i) = \sum_{l=0}^{i} \sum_{j=1}^{\hat{D}^c(l,k)} V_{CHC}(\mathbf{S}^j(\mathbf{p}_{l,k}(j))),$$

where $V_{CHC}(\mathbf{S}^j)$ is the volume of the CHC computed for the box $\mathbf{S}^j = [\mathbf{u}^j, \mathbf{v}^j] \in \mathbb{R}^z \times \mathbb{R}^z$, $\max(z) = 2k$, determined for the j-th c.c.d. of l ones into k groups following the rûle explained in Corollary 4.3.3, *where $\mathbf{p}_{l,k}(j)$ is a $k \times \max(m_s, s = 1, \ldots, k)$ matrix of zeros and ones corresponding to the j-th c.c.d. of l ones into k groups of dimensions m_s, $s = 1, \ldots, k$ and finally accounting for the permutation of the elements into the groups.*

We finally observe that the combinatoric approach explained before is well-defined in principle only for the one-level dependence structure, i.e., not hierarchical copulas. In fact if we have a hierarchy, it means that we suppose the counting event at hand is affected by the presence of a common factor, measured by the dependence between groups and which could be identified as a systemic risk and of several idiosyncratic factors represented by risks attaching thought the dependence within groups. Furthermore the discussed model is essentially based on a not hierarchical dependence structure which prevents us to introduce an explicit discussion about the contagion component of the systemic and idiosyncratic factors. To introduce the contagion effect in the model, we would have a hierarchy whose dependence levels were directly linked to the arrivals of the counting random event and then to the level of contagion itself. In fact we point out that a hierarchical structure affects the arrival of events such that their combinatoric distributions wouldn't be any more equally probable. This argument implies that a pure hierarchical structure can be introduced only if the arrival policy of events is taken into account compatibly with the dependence structure itself. We explained in Chap. 3 how introducing the arrival policy restricting the combinatorial distribution to the distorted ones through a random matrix. Will see in Chap. 5 the detail on how the arrival policy could be introduced into a combinatoric model of this kind.

4.3.7 The Boxes' Definition for the CHC-Volume Computation

As explained in Proposition 4.3.1, the distribution of the counting variable is evaluated as a sequence of copula volumes. Such volumes correspond to special boxes which are detailed in the following MatLab™ code.

```
%This function determines the coordinates of the volume.
%Given the marginals of the groups, pDefSpreadVector, and the row
%corresponding to a c.c.d., finalVector, the code generates two vectors
%whose elements are defined following the rule:
%if the i-th element of the c.c.d. vector is equal to 0, then the i-th
%element of u is the i-th element of pDefSpreadVector and the i-th
%element of v is 1;
%if the i-th element of the c.c.d. vector is equal to 1, then the i-th
%element of u is 0 and the i-th element of v is the i-th element of
%pDefSpreadVector;
function[u,v]= subsSwitch(finalVector,pDefSpreadVector)
for i=1:size(pDefSpreadVector,2)
if finalVector(1,i)≥1
u(1,i)=0;
v(1,i)=pDefSpreadVector(1,i);
else
u(1,i)=pDefSpreadVector(1,i);
v(1,i)=1;
end
end
u=[u(1,:)];
v=[v(1,:)];
```

4.3.8 The Dynamic Version of the Combinatoric-Approach

The new approach presented before is actually static given the static character of the copula tool. In order to overcome this constraint, we need to assume a dynamic evolution of the input features of the clustering. Therefore, given the vector of input $\mathbf{I}(t)$, whose dynamic is

$$\Delta\mathbf{I}(t) = a(t, \mathbf{I}(t))\Delta t + b(t, \mathbf{I}(t))\Delta W(t),$$

where $a(.,.)$, $b(.,.)$ are assigned functions and $\Delta W(t)$ stands for the increment of a Brownian motion from t to $t+1$, we are able to recover the dynamic of the clustering cardinalities, i.e., $\mathcal{C} = \{\mathbf{m}(t), \mathbf{m}(t+1), \ldots\}$. Hence at every t we recover the set of matrix $\mathbf{p}_{l,k}$, $\forall l$, providing the combinatoric compatible distributions of l events in k homogeneous groups, in order to have a set of matrices at every t, i.e., $\mathcal{M}_t = \{\mathbf{p}_{l,k}(t), \forall l \in \sum_{s=1}^{k} m_s(t)\}$ and a set of counting variables, i.e., $\mathcal{R} = \{r(t), \forall t\}$.

The dynamic in discrete time of the counting variable allows to recover a pdf surface $\mathcal{S}_r = \{P_r(l,t), \forall l \in \sum_{s=1}^{k} m_s(t), \forall t\}$ in line with the instruction provided

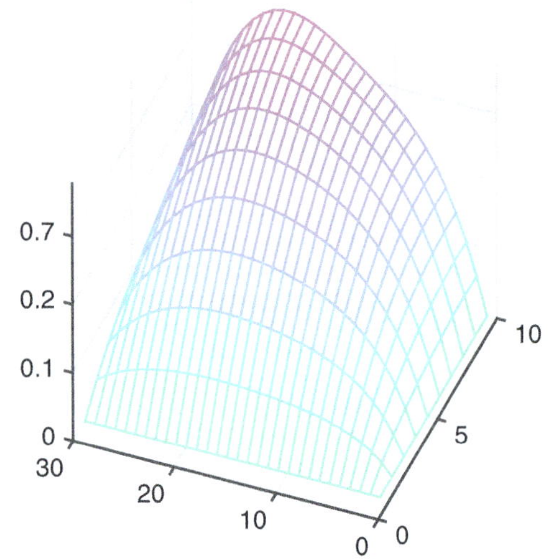

Fig. 4.9 The pdf surface of a counting variable over 10 periods

in Corollary 4.3.3. In Fig. 4.9 is represented a pdf surface[8] of a variable which counts the occurrences of a random event in a set of 30 random variables.

References

1. Bernardi, E., & Romagnoli, S. (2011). Computing the volume of an high-dimensional semi-unsupervised hierarchical copula. *International Journal of Computer Mathematics, 88*(12), 2591–2607.
2. Bernardi, E., & Romagnoli, S. (2013). A clusterized copula-based probability distribution of a counting variable for high-dimensional problems. *The Journal of Credit Risk, 9*(2), 3–26.
3. Bortkiewicz, L. J. (1898). *The Law of Small Numbers*. Leipzig: B.G. Teubner.
4. Cameron, A. C., & Trivedi, P. K. (1998). *Regression Analysis of Count Data*. (Econometric Society Monograph, Vol. 30). Cambridge: Cambridge University Press.
5. Consul, P. C., & Jain, G. C. (1973). A generalization of the Poisson distribution. *Technometrics, 15*(4), 791–799.
6. Davis, M., & Lo, V. (2001). Infectious defaults. *Quantitative Finance, 1*, 382–386.
7. Driessen, J. (2005). Is default event risk priced in corporate bonds? *Review of Financial Studies, 18*, 165–195.
8. Duffee, G. R. (1999). Estimating the price of default risk. *The Review of Financial Studies, 12*(1), 197–226.
9. Duffie, D., & Singleton, K. (1999). Simulating correlated defaults. Stanford University, *working paper*.
10. Elizalde, A. (2005). Credit Risk Models I: Default Correlation in Intensity Models. *working paper*.

[8]The representation has been done smoothing and interpolating the points process, for graphical convenience.

11. Greenwood, M., & Yule, G. U. (1920). An inquiry in to the nature of frequency distributions of multiple happenings, with particular reference to the occurrence of multiple attacks of disease or repeated accidents. *Journal of the Royal Statistical Society, 83*, 255–279.
12. Jarrow, R., & Yu, F. (2001). Counterparty risk and the pricing of defaultable securities. *Journal of Finance, 56*, 1765–1799.
13. Joe, H., & Zhu, R. (2005). Generalized Poisson distribution: The property of mixture of Poisson and comparison with negative binomial distribution. *Biometrical Journal, 47*, 219–229.
14. Jørgensen, B., & De Souza, M. C. P. (1994). Fitting Tweedie's compound Poisson model to insurance claims data. *Scandinavian Actuarial Journal, 1994*, 69–93.
15. Kemp, C. D. (1967). Stuttering-Poisson distributions. *Journal of the Statistical Enquiry of Ireland, 21*(5), 151–157.
16. Lomnitz, C. (1994). *Fundamentals of Earthquake Prediction*. New York: Wiley.
17. Mack, T. (1997). Measuring the variability of chain ladder reserve estimates. *CCasualty Actuarial Society Forum*, Spring 1994.
18. McCullagh, P., & Nelder, J. A. (1989). *Generalized Linear Models*. London: Chapman and Hall.
19. Poisson, S. D. (1837). *Probabilité des jugements en matière criminelle et en matière civile, précédées des règle générales du calcul des probabilitiés*. Paris: Bachelier.
20. Renshaw, A. E., & Verrall, R. J. (1998). A stochastic model underlying the chain-ladder technique. *British Actuarial Journal, 4*, 903–923.
21. Revfeim, K. J. A. (1984). An initial model of the relationship between rainfall events and daily rainfalls. *Journal of Hydrology, 75*, 357–364.
22. Thompson, C. S. (1984). Homogeneity analysis of a rainfall series: An application of the use of a realistic rainfall model. *Journal of Climatology, 4*, 609–619.
23. Tweedie, M. C. K. (1984). An index which distinguishes between some important exponential families. In J. K. Ghosh & J. Roy (Eds.), *Applications and New Directions. Proceeding of the Indian Statistical Golden Jubilee International Conference* (pp. 579–604). Calcutta: Indian Statistical Institute.
24. Wüthrich, M. V. (2003). Claims reserving using Tweedie's compound Poisson model. *Astin Bulletin, 33*(2), 331–346.
25. Zang, F. X. (2003). What did the credit market expect of Argentina default? Evidence from Default Swap Data. *working paper*, Federal Reserve Board, Division of Research and Statistics, Washington.

Chapter 5
A New Copula-Based Approach for Counting: The Distorted and the Limiting Case

Abstract The purpose of this chapter is to examine in depth the new algorithm for counting with dependence, assuming at first to have a hierarchy in the dependence structure and then approaching the problem in very high dimensional cases. The hierarchical structure leads to a set of distorted distributions, selected by a random matrix representing the arrival policy of the random event, while for the tractability of high dimensional problems, one needs to work under approximation assumptions, resulting into a limiting copula-based counting approach.

The goal here is to delve deeper into the new algorithm for counting with dependence, introduced in Chap. 4. As noted before, the technique of distributing the random event in a set of classes is an essential feature of the algorithm. If all the combinatoric distributions are considered, we implicitly assume not to have any hierarchy in the dependence structure, while the set of distorted distributions, selected by a random matrix representing the arrival policy of the random event, is compatible with a hierarchical structure.

Another important question concerns the way one can reduce the complexity of the problem in very high-dimensional cases. When the clustering doesn't allow enough simplification, we may have to recur to approximation results used in the so-called limiting models. This assumption causes us to using the limiting clusterized copulas introduced in Chap. 2.

We'll discuss in the following the distorted model and then do an in-depth analysis of our new approach of counting under the limiting copula assumption.

5.1 The Distorted Copula-Based Approach: Fatal Event

The proposed model is a *distorted* version of the CHC approach for which we point to Chaps. 2 and 4 (see also Bernardi and Romagnoli 4, Bernardi and Romagnoli 6, for further details).

E. Bernardi, S. Romagnoli, *Counting Statistics for Dependent Random Events*,
https://doi.org/10.1007/978-3-030-64250-1_5

CHC approach stands for a combinatoric approach aimed to the calculus of the combinatoric distribution of the occurrences of a random event into a set of groups; each one of these distributions is associated to a volume of the CHC whose aggregation allows us to compute the cdf of the counting variable. All the possible combinatoric distributions, representing all possible scenarios, are considered since we don't assume any arrival policy of the random event, i.e., we assume that the probability that the event affects a variable i is equal to that of variable j, $\forall i, j$. It is this homogeneous policy of arrival that enables us to consider a proper hierarchical dependence structure, whose different levels of dependence are not compatible with a homogeneous distribution of the event among the groups. We may think of an arrival policy as a rule governing the appearance of an event: the arrival looks like the attraction by a magnet with an intensity equal to an assigned probability (that is the probability of event's occurrence) where these probabilities are linked by a dependence structure which is, possibly, hierarchical. A hierarchical structure induces a distortion on the arrival of the event that defines the arrival policy together with the marginals. The main novelty of this approach is to consider the impact of the hierarchy not only on the aggregation step but also on the distribution phase. A particular arrival policy induced by a hierarchy renders a proper hierarchical copula compatible with a combinatoric approach of this kind. We consider here a fatal event as, for example, the default's event.

An arrival policy is formalized through a random matrix A, called *arrival matrix*.[1]

Definition 5.1.1 (Arrival matrix) Given a set of n random variables $\{X^i, i = 1, \ldots, n\}$ representing the default's events, whose marginal probabilities, i.e., the probabilities that the event happens, are $\{p^i, i = 1, \ldots, n\}$, the arrival matrix is a stochastic matrix A, whose components a_{ji} are dependent $0-1$ r.vs such that

$$a_{ji} = \begin{cases} 1 & \text{with probability } p^i \\ 0 & \text{with probability } 1 - p^i \end{cases}$$

and where the dependency is induced by the assumed hierarchy.

We can observe that the number of columns of A corresponds to the dimension of the r.vs set, i.e., n, while the number of rows represents the number of distorted combinatoric distributions or the number of the distributions considered more probable.[2] The arrival policy distorts the combinatoric distributions in such a way that some of these become negligible since the mass of probability is no more

[1]For a basic introduction on random matrix theory, we refer to the brief introductory material of Chap. 3.

[2]The matrix A represents an hidden hierarchy among the variables since it selects only some of the combinatoric distribution, i.e., the distorted ones. This hidden hierarchy represents the effect of the hierarchical structure on the arrival policy.

equally distributed among these but it is concentrated on the subset of the distorted ones.

We can decompose the arrival matrix into several random sub-matrices. In fact each row j of A is associated to an integer $k \in \mathbb{N}$ equal to the sum of the components, i.e., $k = \sum_{i=1}^{n} a_{ji}$. If we select the rows associated to the same integer k, we generate a sub-matrix $\mathbf{p}_{k,n}$ that is a $0-1$ random matrix of the distorted combinatoric distributions of the integer k into n places. Indeed the arrival matrix is the set of all the sub-matrices $\mathbf{p}_{k,n}$, $\forall k$.

5.1.1 From a Not Distorted to a Distorted Structure: A Probabilistic Discussion

CHC approach is a combinatoric algorithm based on the calculus of the combinatoric distribution of a random event e into a set of groups: each one of these combinatoric distributions is then associated to a scenario, whose probability to happen corresponds to a volume of the CHC. More formally, for a set of n one-dimensional groups, the set of all the combinatoric distributions of the random event e corresponds to a 0-1 matrix, i.e.,

$$\mathbf{E} = \begin{pmatrix} 0\,0 \ldots\ldots 0 \\ 1\,0 \ldots\ldots 0 \\ \vdots\ \vdots\ \vdots\ \ \vdots\ \ \vdots \\ 1\,1 \ldots\ldots 1 \end{pmatrix}$$

whose dimension is $2^n \times n$. Every row of $\mathbf{E}$ represents a possible scenario, whose probability to happen is given by a volume of the copula function representing the dependence structure of groups. The volume corresponding to the jth row of matrix $\mathbf{E}$ is specified by a box $\mathbf{S}^j = [\mathbf{u}^j, \mathbf{v}^j] \in \mathbb{R}^2 \times \mathbb{R}^2$, determined following the rule:

- if $E_{ji} = 0$, $u_i^j = p_i$ and $v_i^j = 1$ for $i = 1, \ldots, n$, while if $E_{ji} = 1$, $u_i^j = 0$ and $v_i^j = p_i$ for $i = 1, \ldots, n$,

where $\{p_i, i = 1, \ldots, n\}$ are the probabilities to happen of the event e while E_{ji} stands for the (j, i)-th entry of matrix $\mathbf{E}$.

The aggregation of these volumes allows us to compute the cdf of the counting variable. The total set of the combinatoric distributions, representing every possible scenario, is considered, since the CHC approach does not account for any arrival policy of the event, i.e., we assume that the probability the event affects a variable i is the same for every $i = 1, \ldots, n$. It is this homogeneous assumption about the arrivals that prevents us from considering a proper hierarchical dependence structure, whose different levels of dependence are not compatible with equally probable combinatoric distributions of the event.

Let us consider the arrival matrix defined before. The arrival policy distorts the combinatoric distributions in **E** in such a way that some of these become negligible since the mass of probability is no more equally distributed among these but it is concentrated on the subset of the distorted ones. From a methodological point of view, to construct the matrix **A**, we need to sample from the copula function representing the dependence structure among variables. In the case of proper n-dimensional Archimedean copulas, we know that the generator ϕ must be completely monotone and $\phi = \mathcal{LS}[F]$, where the Laplace-Stieltjes transform of F is defined as $\mathcal{LS}[F](t) = \int_0^{+\infty} e^{-tx} dF(x), t \in [0, +\infty)$. The link between completely monotone functions and Laplace-Stieltjes transforms constitutes the building block of an efficient sampling algorithms for (partially) exchangeable and not exchangeable Archimedean[3] copulas based on a mixture representation of the hierarchical copula in term of functions $F_{n-1} = \mathcal{LS}^{-1}[\phi_{n-1}]$ and $F_{ij} = \mathcal{LS}^{-1}[\phi_{ij}(V_{n-1})]$, where $\phi_{ij}(t, x) = e^{-x\phi_i^{-1}(\phi_j(t))}, t \in [0, +\infty), x \in (0, +\infty)$, and where $\{\phi_{n-1}, \phi_{n-2}, \ldots, \phi_1\}$ stands for the set of generators ordered starting from the highest aggregation's level (which represents the root copula), while $\phi_{i,j} = \phi_i^{-1} \circ \phi_j$ represents the compound operator useful for the aggregation of parent i and child j, $\forall i, j$. We refer to the following algorithm, whose first four steps have been developed in Hofert and Mächler [12], in the case of nested Archimedean copulas.

Algorithm 5.1.1 *Let C be a nested Archimedean copula with root copula C_{n-1}, generated by ϕ_{n-1}, and* **U** *be a vector of the same dimension as C. We follow the next steps:*

1. *Sample $V_{n-1} \sim F_{n-1} = \mathcal{LS}^{-1}[\phi_{n-1}]$*
2. *For all components of C_{n-1} that are nested Archimedean copulas do:*
   - *set C_{n-2} with generator ϕ_{n-2} to the nested Archimedean copula u*
   - *sample $V_{n-1,n-2} \sim F_{n-1,n-2} = \mathcal{LS}^{-1}[\phi_{n-1,n-2}(V_{n-1})]$*
   - *set $C_{n-1} = C_{n-2}$, $\phi_{n-1} = \phi_{n-2}$, and $V_{n-1} = V_{n-1,n-2}$ and continue with* (2)
3. *For all other components u of C_{n-1} do:*
   - *sample $X \sim U[0, 1]$*
   - *set the component* **U**, *corresponding to u to $\phi_{n-1}\left(\frac{-\log X}{V_{n-1}}\right)$*
4. *Return* **U**
5. *Given* **U**, *we define vector A_j of the same dimension whose ith entry is $a_{ji} = 0$ if $U_i \leq 1 - p_i$ and $a_{ji} = 1$ if $U_i > p_i$, where p_i is the ith margin, while U_i stands for the ith entry of* **U**.
6. *Return A_j and repeat 2^n times from* (1).

As we can observe, the basic idea of the previous algorithm is to sample 2^n times from the common dependence structure and transform the obtained variates to the

[3] See Chap. 2 for details on the hierarchical Archimedean class of copula functions.

proper margins a_{ji}. In particular we have to transform margins into binomial. Doing so we are able to create the arrival matrix $\mathbf{A}$, whose jth row is given by vector A_j. Then given a distortion function $\mathcal{D}$ (represented by the Algorithm 5.1.1) which is a function of the set of generators $\{\phi_i, i = 1, \ldots, n-1\}$ and the set of margins $\{p_1, i = 1 \ldots, n\}$, we can represent the arrival matrix as a distorted version of matrix $\mathbf{E}$, i.e., $\mathbf{A} = \mathcal{D}(\mathbf{E})$.

5.1.2 Distorted Copula-Based Distribution of a Fatal Counting Variable

The proposed procedure involves the computation of the probabilities assigned to the distorted combinatoric distributions, representing the most likely scenarios, given an arrival policy of the fatal random event as, for example, the default's event. These probabilities are evaluated as volumes of the copula that formalizes the dependence structure among variables and that we assume to be hierarchical Archimedean (and then distorted by the arrival policy induced by the hierarchy itself). We name this kind of clusterized hierarchical copula restricted to the distorted distribution as *distorted hierarchical copula*, and we note it as DHC in the following. In the same spirit of the representation of the volume of a CHC (see Bernardi and Romagnoli 2), where the whole set of the combinatorial distributions of the random event is included into the computation and is assumed to be equally probable, the computation of the volume of a DHC must be restricted to the distorted combinatorial distributions induced by the arrival policy. The following definition generalizes the case of homogeneous arrival policy assigning the same probability to every scenario, depicted in Chap. 4.

Proposition 5.1.2 (DHC volume) *Given a hierarchical Archimedean copula function of dimension n and an arrival matrix $\mathbf{A} = \{\mathbf{p}_{k,n}, k \in \mathbb{N}, k \leq n\}$, the volume of the DHC defined by the n-dimensional box $\mathbf{S} = [\mathbf{u}, \mathbf{v}]$ with $\mathbf{u}, \mathbf{v} \in [0,1]^n$, $\mathbf{u} \leq \mathbf{v}$ may be represented as*

$$V_{DHC}(\mathbf{S}) = \sum_{k=0}^{n} (-1)^k \sum_{j=1}^{D^d(k,n)} DHC(\mathbf{c}(\mathbf{p}_{k,n}(j)))$$

where $D^d(k,n)$ denotes the number of distorted combinatorial distributions (d.c.ds in the following)[4] *of the integer k into n places, $\mathbf{p}_{k,n}(j)$ is the j-th row of the sub-arrival matrix $\mathbf{p}_{k,n}$ whose dimension is $D^d(k,n) \times n$, $\mathbf{c}(\mathbf{p}_{k,n}(j))$ is a vector of dimension n such that $c_{j,i} = v_i$ if $p_{j,i} = 0$ and $c_{j,i} = u_i$ if $p_{j,i} = 1$, where*

[4] We refer to $D^d(k,n)$ as the number of the combinatorial distributions in the corresponding sub-arrival matrix $\mathbf{p}_{k,n}$, i.e., the most probable ways to distribute the integer k into n places.

u_i, v_i are the i-th element of the corresponding vector and $c_{j,i}$ denotes the (j, i)-th element of the corresponding vector and $DHC(\mathbf{c}(\mathbf{p}_{k,n}(j)))$ is the hierarchical Archimedean copula computed for the j-th d.c.d. of the sub-arrival matrix $\mathbf{p}_{k,n}$.

Remark 5.1.1 We observe that a hierarchical distorted copula function is a hierarchical copula function where we consider only the distribution of events which are the most probable, based on the arrival matrix A. This copula takes into account the twofold impact of the hierarchy: the impact on the aggregation (in the calculation of the copula-volume) and on the distribution phase.

Example 5.1.1 (DHC volume) Given a set of three variables, placed in single groups and an arrival matrix $\mathbf{A} = \{\mathbf{p}_{k,3}, k \in \mathbb{N}, k \leq 3\}$ such that

$$\mathbf{p}_{0,3} = [0\,0\,0] \quad \mathbf{p}_{1,3} = [1\,0\,0]$$

$$\mathbf{p}_{2,3} = \begin{bmatrix} 1\,1\,0 \\ 0\,1\,1 \end{bmatrix} \quad \mathbf{p}_{3,3} = [1\ 1\ 1],$$

we compute the DHC volume defined by the three-dimensional box $\mathbf{S} = [\mathbf{u}, \mathbf{v}]$. Following the rule described in Proposition 5.1.2, we have

$$DHC_0 = DHC(v_1, v_2, v_3), \quad DHC_1 = DHC(u_1, v_2, v_3),$$

$$DHC_2 = DHC(u_1, u_2, v_3) + DHC(v_1, u_2, u_3), \quad DHC_3 = DHC(u_1, u_2, u_3).$$

We observe that the assumed arrival matrix assigns a null probability to some scenarios as, for example, to $[1\ 0\ 1]$ that doesn't enter in $\mathbf{p}_{2,3}$. Then finally $V_{DHC}(\mathbf{S}) = \sum_{k=0}^{3}(-1)^k DHC_k$ that is represented in Fig. 5.1, allowing to understand the intuition behind volume's computation.

As pointed out before, our approach is copula-based in the sense that, given a copula function representing the dependence structure of a given basket of variables, we consider an event which may involve such variables and define a counting variable on this event.

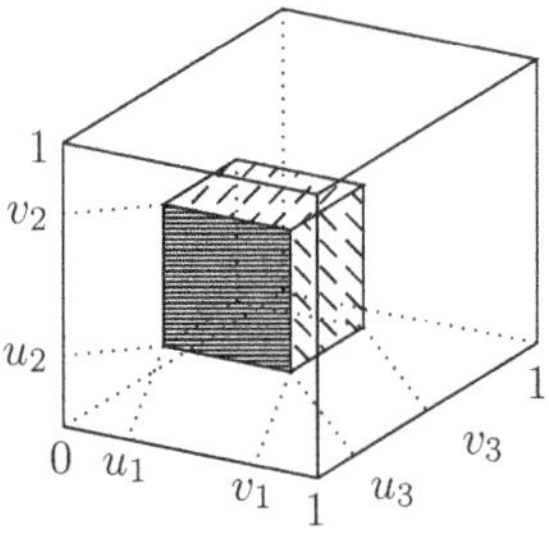

Fig. 5.1 Graphical representation of $V_{DHC}(\mathbf{S})$

Definition 5.1.2 (Counting variable linked to a n-dimensional DHC) In the same setting of Proposition 5.1.2, the counting variable linked to the copula function $DHC(\mathbf{c}(\mathbf{p}_{k,n}(j)))$ is given by

$$r(j) = \sharp(\mathbf{p}_{k,n}(j))$$

which counts the number of elements equal to one in the j-th row of the sub-arrival matrix $\mathbf{p}_{k,n}$. Clearly we have $r(j) = k, \forall j$.

Our aim is to recover the probability distribution of the counting variable linked to the n-dimensional copula function. To achieve this we observe that a number k corresponds, through its probability function, to a value in $[0, 1]$ by considering the d.c.ds of k ones into n places (we enumerate this d.c.d. with the index j) induced by the arrival matrix. The j-th d.c.d. of the sub-matrix $\mathbf{p}_{k,n}$ generates a pair of coordinates,[5] i.e., the j-th d.c.d. generates the box $\mathbf{S}^j$ where we compute the volume. The sum of the volumes computed for all the coordinates generated by the d.c.ds of $\mathbf{p}_{k,n}$ represents the probability to count the most likely k, i.e., the most probable scenarios, given an arrival hypothesis A.

Corollary 5.1.3 (Probability distribution of the counting variable linked to a n-dimensional DHC) *In the same setting of Proposition* 5.1.2, *the probability distribution of the counting variable linked to the n-dimensional copula DHC of dimension n, with marginals* $\mathbf{u} \in [0, 1]^n$ *and arrival matrix A, is the function* $P_r : [0, n] \to [0, 1]$ *such that*

$$P_r(k) = \sum_{j=1}^{D^d(k,n)} V_{DHC}(\mathbf{S}^j(\mathbf{p}_{k,n}(j)))$$

where $\mathbf{p}_{k,n}(j)$ *is the j-th row of the sub-arrival matrix* $\mathbf{p}_{k,n} \in M(D^d(k, n) \times n)$ *representing the j-th d.c.d. of k ones into n places and where* $V_{DHC}(\mathbf{S}^j)$ *is the volume of the distorted hierarchical copula computed for the box* $\mathbf{S}^j = [\mathbf{u}^j, \mathbf{v}^j] \in \mathbb{R}^n \times \mathbb{R}^n$, *determined for the j-th d.c.d. of k ones into n places following the rule:*

- *if* $p_{j,i} = 0, u_{j,i} = u_i$ *and* $v_{j,i} = 1$ *for* $i = 1, \ldots, n$ *and* $j = 1, \ldots, D^d(k, n)$;
- *if* $p_{j,i} = 1, u_{j,i} = 0$ *and* $v_{j,i} = u_i$ *for* $i = 1, \ldots, n$ *and* $j = 1, \ldots, D^d(k, n)$;

Referring to the evaluation of the probability distribution of the counting variable explained in Proposition 5.1.2, in the following example, we lay out its computational phase.

Example 5.1.2 (Pdf computation.) From Proposition 5.1.2, we know that to compute the probability distribution of the counting variable linked to a n-dimensional DHC, we must first compute the d.c.ds of all integers $k \leq n$. In fact we need to

[5] The rule which explains how the coordinates are generated, is presented in the following Corollary.

consider the d.c.ds of all the integers starting from zero to n, generating the sub-matrices $\mathbf{p}_{k,n}$, $k \in \mathbb{N}$. Given an integer and then a sub-arrival matrix, each d.c.d. corresponds to a box, i.e., a pair of coordinates, whose volume is computed as detailed in Corollary 5.1.3.

For example, if we refer to the sub-arrival matrix $\mathbf{p}_{2,3}$ of example 5.1.1, corresponding to the most probable scenarios which distribute 2 random events into a set of 3 variables and we call the d.c.ds following the row's order, the coordinates defining the box $\mathbf{S}^1$ and corresponding to the first row of $\mathbf{p}_{2,3}$, are:

$$\mathbf{u}_1 = \begin{bmatrix} 0 \; 0 \; u_3 \end{bmatrix}, \quad \mathbf{v}_1 = \begin{bmatrix} u_1 \; u_2 \; 1 \end{bmatrix},$$

while the coordinates defining the box $\mathbf{S}^2$ and corresponding to the second row of $\mathbf{p}_{2,3}$, are:

$$\mathbf{u}_2 = \begin{bmatrix} u_1 \; 0 \; 0 \end{bmatrix}, \quad \mathbf{v}_2 = \begin{bmatrix} 1 \; u_2 \; u_3 \end{bmatrix},$$

where $\begin{bmatrix} u_1 \; u_2 \; u_3 \end{bmatrix}$ are the marginals, i.e., the occurrence probabilities of the event (as, e.g., the default probabilities) for the corresponding variables. In Fig. 5.2 the volumes defined by these coordinates, are shown. Finally we observe that the sum of the volumes computed for every row of the same sub-arrival matrix, as, for example, of $\mathbf{p}_{2,3}$, represents the distorted probability that the counting variable is equal to the integer corresponding to this matrix, i.e., $P_r(2) = V_{DHC}(\mathbf{S}^1) + V_{DHC}(\mathbf{S}^2)$.

Corollary 5.1.4 (Cumulated probability distribution of the counting variable linked to a n-dimensional DHC) *In the same setting of Proposition* 5.1.2 *and Corollary* 5.1.3, *the cumulative probability distribution of the counting variable linked to the n-dimensional copula DHC of dimension n and with marginals* $\mathbf{u} \in [0,1]^n$ *is the function* $F_r : [0, n] \to [0, 1]$ *such that*

$$F_r(i) = \sum_{k=0}^{i} \sum_{j=1}^{D^d(k,n)} V_{DHC}(\mathbf{S}^j(\mathbf{p}_{k,n}(j)))$$

where $V_{DHC}(\mathbf{S}^j)$ *is the volume of the distorted hierarchical copula computed for the box* $\mathbf{S}^j = [\mathbf{u}^j, \mathbf{v}^j] \in \mathbb{R}^n \times \mathbb{R}^n$, *determined for the* j*-th d.c.d. of* k *ones into* n *places following the rule explained in Corollary* 5.1.3, *where* $\mathbf{p}_{k,n}(j)$ *is the* j*-th row*

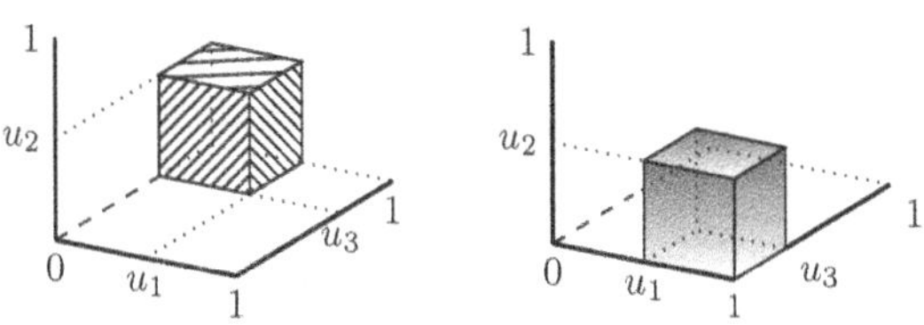

Fig. 5.2 Graphical representation of $V_{DHC}(\mathbf{S}^1)$ and $V_{DHC}(\mathbf{S}^2)$

of the sub-arrival matrix $\mathbf{p}_{k,n} \in M(D^d(k,n) \times n)$ *corresponding to the* j*-th d.c.d. of* k *ones into* n *places.*

Referring to the evaluation of the probability distribution of the counting variable explained in Corollary 5.1.3, in the following example, we lay out the computational phase of the coordinates defining the box.

Example 5.1.3 (Box's definition) To compute the probability distribution of the counting variable linked to a copula function, we must first compute the d.c.ds of an integer. We must consider the d.c.ds for all the integers from zero to the dimension of the problem generating the sub-matrices $\mathbf{p}_{k,n}$, $k \in \mathbb{N}$. Given an integer and then a sub-arrival matrix, each d.c.d. corresponds to a box, i.e., a pair of coordinates, where we must compute the volume as explained in Corollary 5.1.3.

For example, if the second row of the sub-arrival matrix $\mathbf{p}_{2,3}$ for the integer 2 into three places is the following:

$$[0\ 1\ 1]$$

and we call the d.c.ds following the row's order,[6] the corresponding coordinates defining the box $\mathbf{S}^2$ are:

$$\mathbf{u}^2 = \begin{bmatrix} v_1 & 0 & 0 \end{bmatrix}$$
$$\mathbf{v}^2 = \begin{bmatrix} 1 & v_2 & v_3 \end{bmatrix}$$

where $\begin{bmatrix} v_1 & v_2 & v_3 \end{bmatrix}$ are the marginals, i.e., the event's probabilities (as, e.g., the default probabilities) for the corresponding variables. Finally we observe that the sum of the volumes computed for all rows of the same sub-arrival matrix, i.e., considering the d.c.ds of the same integer k, represents the distorted probability that the counting variable is equal to k itself. The distortion is induced by the arrival policy and then by the marginals and the hierarchy of variables.

5.2 The Distorted Copula-Based Approach: Not Fatal Event

In this section we discuss the distorted model in the case of not fatal event instead of a fatal one, as, for example, the percentage of losses or the number of claims. For this reason, to make the problem compatible with a combinatoric approach similar to that proposed in the DHC model, we choose a percentage unit x and define the number of percentage units, $n_x = \frac{100}{x}$, that will be distributed over the underlying variables to recover the cdf of the counting variable. In line with Bernardi and Romagnoli [5], here we consider a portfolio that we assume clusterized

[6] Following this calling order, the d.c.d. considered here is the second one, i.e., $j = 2$.

into k classes. Moreover these classes may be also clusterized at a second level such to identify the *within* dependence structure. Each second level sub-group is characterized by its dimension $n_i^u, i = 1, \ldots, k, u = 1, \ldots d_i$ where d_i denotes the number of sub-groups in class $i, i = 1, \ldots k$ and its centroids whose marginal probabilities are $\mathbf{p}^{(i,u)}, i = 1, \ldots, k, u = 1, \ldots d_i$ whose z-th component $p_z^{(i,u)}$ is the probability to have a percentage of not fatal event equal to zx into the u-th sub-group of class i.

The idea is to represent the arrival policy of events assuming to have a clusterized portfolio whose dependence structure is hierarchical and then represented by a dependence *within* and *between* classes, i.e., it is a clusterized hierarchical copula (CHY for short) restricted to the distorted distributions. We assume that the arrival of events is determined by idiosyncratic factors that correspond to the marginal distortion induced by the marginal probabilities and by a systemic factor. Obviously the systemic risk can be decomposed into a proper market risk and a contagion factor but we know that it is very difficult to distinguish them; in our model we assume that the *between* group dependence structure is the expression of market risk, while the *within* dependence stands for the contagion one. Furthermore in our model we are able to assume several degrees of contagion depending on the cardinality of the second level clustering result, i.e., on the number of emerging sub-groups $d_i, i = 1, \ldots, k$; the greater is d_i and the lower is the contagion degree in group i. In fact each sub-group is assumed to be homogeneous and then perfectly correlated with the centroid; this implies that the larger is the sub-group and the greater is the number of elements that move all together. If classes are not clusterized into sub-groups, we have total contagion inside each class,[7] while if a second level of grouping is considered, we have total contagion inside sub-groups but partial contagion inside classes whose level depends on the cardinalities of sub-groups themselves. On the other hand, the case of no contagion at all is reached only with single sub-groups that are linked by a one-level dependence structure like 1-CHC. In this limit case, obviously we lose the hierarchy because all the underlyings are affected by the same systemic risk that is totally explained by the market component of risk.

Given a perfect correlation assumption between centroids and all the components of each sub-group in line with the previous arguments, we consequently reduce the dimension of the problem considering an arrival policy that concerns only the centroids of sub-groups. An arrival policy is formalized through a random matrix $\mathbf{A}$, called *arrival matrix*.

Definition 5.2.1 (Arrival matrix) Let be given a set of random vectors $\{\mathbf{X}^i, i = 1, \ldots, k\}$ where $\mathbf{X}^i$ is a vector of dimension $n_x d_i$ representing the events' arrivals on sub-groups' centroids of class i, whose marginal probabilities, i.e., the probabilities that the event happens, are $\{\mathbf{p}^i, i = 1, \ldots, k\}$, where $\mathbf{p}^i = (\mathbf{p}^{(i,u)}, u = 1, \ldots, d_i)$

[7]In this case we refer to a special case of CHC copula, i.e., a perfectly homogeneous clusterized copula, restricted to the distorted distributions.

and each $\mathbf{p}^{(i,u)}$ is a vector whose z-th component is the probability to have one more unit of event after having already suffered $z-1$ percentage units.[8] An arrival matrix is a stochastic matrix A, whose components $\mathbf{a}_{ji}$ is a stochastic vector row $\mathbf{a}_{ji} = (\mathbf{a}_{ji}^{(u)}, u = 1, \ldots, d_i)$ whose z-th component of $\mathbf{a}_{ji}^{(u)}$ that represents the arrival of z-th incremental unit of event into u-th sub-group of class i is a dependent $0-1$ r.v.

$$a_{ji,z}^{(u)} = \begin{cases} 1 & \text{with probability } p_z^{(i,u)} \\ 0 & \text{with probability } 1 - p_z^{(i,u)} \end{cases}$$

and where the dependency is induced by the assumed hierarchy.

Given an arrival matrix $\mathbf{A}$, we must reduce it to a matrix that registers only the arrival of the maximum number of events into each sub-group and for each scenario. This way we recover the *reduced arrival matrix*, representing the scenarios of the aggregated events, whose marginal probabilities are summarized through aggregation, into the so-called aggregated probability matrix.

Definition 5.2.2 (Reduced arrival matrix and aggregated probability matrix) Given an arrival matrix $\mathbf{A}$, the reduced arrival matrix $\bar{\mathbf{A}}$ is a stochastic matrix, whose component $\bar{\mathbf{a}}_{ji}$ is a stochastic row vector $\bar{\mathbf{a}}_{ji} = (\bar{a}_{ji}^{(u)}, u = 1, \ldots, d_i)$ such that

$$\bar{a}_{ji}^{(u)} = \begin{cases} 1 & \text{if } \sum_{z=1}^{n_x} a_{ji,z}^{(u)} \geq 1 \\ 0 & \text{elsewhere} \end{cases}$$

$\bar{\mathbf{A}}$ is associated to the aggregated probability matrix $\bar{\mathbf{P}}$ whose components are the stochastic row vectors $\bar{\mathbf{p}}_{ji} = (\bar{p}_{ji}^{(u)}, u = 1, \ldots, d_i)$ such that $\bar{p}_{ji}^{(u)} = \sum_{z=1}^{m^{(i,u)}} p_z^{(i,u)}$, where $m^{(i,u)} = \max\{z; a_{ji,z}^{(u)} = 1\}$.

We observe that $\bar{\mathbf{P}}$ collects the aggregated probabilities which will be useful in the following because for the definition of copula function, marginals must be uniform rvs. Moreover we point out that the dimension's reduction of the arrival matrix acts into the columns' dimension of $\bar{\mathbf{A}}$ that is now reduced to $\sum_{i=1}^{k} d_i$. To our aim, each scenario and then each row of $\bar{\mathbf{A}}$ must be associated to the information concerning the total number of events and eventually to the monetary value of the involved consequence of the happened event. If we assume that the event corresponds to a percentage of loss, the consequence is the monetary value of the total loss, which is defined below.

[8] We work here with incremental conditional probabilities, i.e., $p_z^{(i,u)} = \mathbb{P}((z-1)x < L^{(i,u)} \leq zx | L^{(i,u)} = (z-1)x)$.

Definition 5.2.3 (Total loss and percentage cardinality vectors) The reduced arrival matrix is associated to the total loss and the percentage cardinality vector. The total loss vector **l**, gives the monetary value of total loss, i.e.,

$$l_j = \frac{\sum_{i=1}^{k} \sum_{u=1}^{d_i} c^{(i,u)} n^{(i,u)} m^{(i,u)}}{V},$$

where $\mathbf{n}^{(i)} = (n^{(i,u)}, u = 1, \ldots, d_i)$ is the vector of sub-groups' cardinalities of class i, $\mathbf{c}^{(i)} = (c^{(i,u)}, u = 1, \ldots, d_i)$ is the vector of sub-groups' centroids of class i, and V is the present value of the whole portfolio.
Further the percentage cardinality **pc** counts for each scenario the number of percentage units involved, i.e.,

$$pc_j = \sum_{i=1}^{k} \sum_{u=1}^{d_i} m^{(i,u)}.$$

We point out that the knowledge of the total loss induces a total order among the scenarios represented by matrix $\bar{\mathbf{A}}$. To this purpose we decompose the reduced arrival matrix into several random sub-matrices $\{\bar{\mathbf{A}}^{(s)}, s = 1, \ldots, d_l\}$; for each $\bar{\mathbf{A}}^{(s)}$ we can say that it is composed by rows associated to several scenarios characterized by the same total loss l_s.[9] Given d_l that is the number of such sub-matrices and the set $\{\bar{\mathbf{A}}^{(s)}, s = 1, \ldots, d_l\}$, we define the linked set of percentage cardinalities $\{\mathbf{pc}^{(s)}, s = 1, \ldots, d_l\}$, whose elements represent the number of percentage losses of each row of $\bar{\mathbf{A}}^{(s)}$.

5.2.1 *The Distorted Copula-Based Distribution of a Not Fatal Counting Variable*

In this section we provide the calculus of the cdf of the counting variable described before, that is, based on the combinatoric distributions distorted by the idiosyncratic and the systemic factors. First of all as our procedure involves the volume of a distorted hierarchical copula function in $\sum_{i=1}^{k} d_i$ dimensions, we propose here a variant of the DHC-volume reported in Proposition 5.1.2 for a fatal event.

Proposition 5.2.1 (Hierarchical Distorted copula volume) *Given a set of sub-reduced arrival matrices* $\{\bar{\mathbf{A}}^{(s)}, s = 1, \ldots, d_l\}$ *and a hierarchical copula function*

[9]We observe that the column's dimension of $\bar{\mathbf{A}}^{(s)}$ is equal to that of $\bar{\mathbf{A}}$ but the row's dimension is stochastic.

DHC of dimension $\sum_{i=1}^{k} d_i$, *the volume of the box* $\mathbf{S} = [\mathbf{u}, \mathbf{v}]$ *with* $\mathbf{u}, \mathbf{v} \in [0, 1]^{\sum_{i=1}^{k} d_i}$, $\mathbf{u} \leq \mathbf{v}$, *may be represented as*

$$V_{DHC}(\mathbf{S}) = \sum_{s=1}^{d_l} \sum_{j=1}^{dim(\bar{\mathbf{A}}^{(s)})} (-1)^{pc_j^{(s)}} DHC(\mathbf{c}(\bar{\mathbf{A}}_j^{(s)}))$$

where $dim(\bar{\mathbf{A}}^{(s)})$ *states for the row's dimension of matrix* $\bar{\mathbf{A}}^{(s)}$, $\bar{\mathbf{A}}_j^{(s)}$ *denotes the* j*-th row of matrix* $\bar{\mathbf{A}}^{(s)}$, *and* $\mathbf{c}(\bar{\mathbf{A}}^{(s)})$ *is a vector of dimension* $\sum_{i=1}^{k} d_i$ *such that* $c_{ji} = v_i$ *if* $\bar{\mathbf{A}}_{ji}^{(s)} = 0$ *and* $c_{ji} = u_i$ *if* $\bar{\mathbf{A}}_{j,i}^{(s)} = 1$, *where* u_i, v_i *are the* i*-th element of the corresponding vector and* c_{ji} *denotes the* (j, i)*-th element of the corresponding matrix and* $DHC(\mathbf{c}(\bar{\mathbf{A}}_j^{(s)}))$ *is the hierarchical copula computed for the* j*-th row of* $\bar{\mathbf{A}}^{(s)}$.

As pointed out before, here we refer to a not fatal event, as, for example, the percentage of loss. In this case we are able to define the corresponding loss variable.

Definition 5.2.4 (Counting variable and loss variable linked to a copula function) In the same setting of Proposition 5.2.1, the counting variable linked to the copula function $DHC(\mathbf{c}(\bar{\mathbf{A}}_j^{(s)}))$ is given by

$$cr(j) = \sharp(\bar{\mathbf{A}}_j^{(s)}),$$

which counts the number of percentage loss's units defining the j-th row of the sub-arrival matrix $\bar{\mathbf{A}}^{(s)}$. Clearly we have $r(j) = pc_j, \forall j$, i.e., it corresponds to the percentage cardinality.

On the other hand, the loss variable linked to the copula $DHC(\mathbf{c}(\bar{\mathbf{A}}_j^{(s)}))$ is given by

$$r(j) = l_s, \forall j$$

and then corresponds to the total loss characterizing the sub-matrix $\bar{\mathbf{A}}^{(s)}$.

Our aim is to recover the probability distribution of the loss variable that is the monetary expression of the counting variable linked to a given hierarchical copula. To achieve it we recall that the total loss l_s corresponds, through its probability function, to a number in [0, 1] by considering the distribution of percentage losses induced by the sub-arrival matrix $\bar{\mathbf{A}}^{(s)}$. The j-th row of $\bar{\mathbf{A}}^{(s)}$ generates a pair of coordinates,[10] defining the box $\mathbf{S}^j$ for which we compute the volume. The weighted

[10] The rule which explains how the coordinates are generated is presented in the following Corollary.

mean[11] of the volumes computed for all the coordinates generated by the rows of $\bar{\mathbf{A}}^{(s)}$ represents the probability to have a mean total loss l_s, given an arrival hypothesis, i.e., given $\bar{\mathbf{A}}$.

Corollary 5.2.2 (Probability distribution of the total loss) *In the same setting of Proposition 5.2.1, the probability distribution of the total loss variable linked to the copula DHC of dimension $\sum_{i=1}^{k} d_i$, with marginals $\mathbf{u} \in [0,1]^{\sum_{i=1}^{k} d_i}$ and reduced arrival matrix $\bar{\mathbf{A}}$, is the function $P_r : \mathcal{R}_l \to [0,1]$ such that*

$$P_r(l_s) = \mathbb{E}(V_{DHC}(\mathbf{S}^j(\bar{\mathbf{A}}_j^{(s)}))f_s,$$

where $\mathbb{E}(.)$ stands for the expected value operator, $\mathcal{R}_l$ is the range of the total loss variable, $f_s = \frac{dim(\bar{\mathbf{A}}^{(s)})}{dim(\bar{\mathbf{A}})}$ is the frequency of l_s, $\bar{\mathbf{A}}_j^{(s)}$ is the j-th row of the sub-arrival matrix $\bar{\mathbf{A}}^{(s)} \in M(dim(\bar{\mathbf{A}}^{(s)}) \times \sum_{i=1}^{k} d_i)$, and $V_{DHC}(\mathbf{S}^j)$ is the volume of the distorted hierarchical copula computed for the box $\mathbf{S}^j = [\mathbf{u}^j, \mathbf{v}^j] \in \mathbb{R}^{\sum_{i=1}^{k} d_i} \times \mathbb{R}^{\sum_{i=1}^{k} d_i}$, determined for the j-th row of $\bar{\mathbf{A}}^{(s)}$ and $\forall j = 1, \ldots, dim(\bar{\mathbf{A}}^{(s)})$ following the rule:

- *if $\bar{\mathbf{A}}_{ji}^{(s)} = 0$, $u_{ji} = u_i$ and $v_{ji} = 1$ for $i = 1, \ldots, \sum_{i=1}^{k} d_i$;*
- *if $\bar{\mathbf{A}}_{ji}^{(s)} = 1$, $u_{ji} = 0$ and $v_{ji} = u_i$ for $i = 1, \ldots, \sum_{i=1}^{k} d_i$.*

Remark 5.2.1 The expectation operator in Corollary 5.2.2 could be thought as the expression of specified subjective beliefs on the scenarios represented by $\bar{\mathbf{A}}^{(s)}$. If the scenarios are assumed to be equally probable, the expectation simply becomes the arithmetic mean operator.

Corollary 5.2.3 (Cumulated probability distribution of the total loss variable) *In the same setting of Proposition 5.2.1 and Corollary 5.2.2, the cumulative probability distribution of the total loss variable linked to the $\sum_{i=1}^{k} d_i$-dimensional copula DHC and with marginals $\mathbf{u} \in [0,1]^{\sum_{i=1}^{k} d_i}$ is the function $F_r : \mathcal{R}_l \to [0,1]$ such that*

$$F_r(l_i) = \sum_{s=0}^{i} P_r(l_s)$$

where $l_0 = 0$ and $P_r(l_s)$ is the probability to have a mean total loss equal to l_s.

Example 5.2.1 (Structural and marginal distortion in DHC model: a credit risk experiment) Our aim is to distinguish between the impact of the dependence structure and of the marginals in the distortion of the distributions. Therefore we consider a portfolio composed by credit exposed assets (i.e., a loan portfolio) for

[11] We talk about weighted mean since we ponderate for the frequency of each scenario with respect to the total number of trials.

which we assume that the event of default occurs if the losses are greater than a fixed level. The portfolio manager evaluates the marginal default's probabilities of each loan and, for structured finance purposes, is interested on the probability distribution of the number of defaults occurring, i.e., on the *loss distribution*. Therefore we consider a fatal event.

In this example we consider a portfolio composed by 10 loans with the following marginal probabilities of default:

$$\mathbf{p} = \left[0.02\ 0.1\ 0.5\ 0.005\ 0.7\ 0.3\ 0.2\ 0.5\ 0.08\ 0.06\right]$$

We consider a dependence structure at the first level represented by a Kendall tau $\tau^1 = 0.048$ and a second level vector of dependence representing the dependence among the ordered couple of variables, i.e.,

$$\boldsymbol{\tau}^{\mathbf{2}} = \left[0.8\ 0.6\ 0.43\ 0.8\ 0.6\right]$$

We consider the following dependence structure:

1. One-level dependence structure, represented by a copula function with τ^1
2. Hierarchical structure, with two levels represented by τ^1 and $\boldsymbol{\tau}^{\mathbf{2}}$
3. Nested structure, where the aggregating dependence level are given by $\boldsymbol{\tau}^{\mathbf{2}}$ decreasingly ordered

We observe that the dependence structure represents the relationships between the defaults; subsequently it impacts their arrivals through the arrival matrix A. This matrix defines the distorted combinatoric distribution whose probability is recovered through the assumed dependence structure. Indeed it is evident that the impact of the dependence structure is twofold since it enters into the determination of the arrival matrix and finally acts into the aggregating step. If we assume independence, i.e., we assume that the arrivals of defaults are independent of the connections among the other variables, the arrival matrix will be distorted only by the marginals probabilities which attract the defaults with different forces. In fact the marginal probabilities of default contribute to determine the arrival matrix along with the dependence structure; to identify the contribution of the dependence structure, we consider the case of homogeneous marginals equal to the centroid of the class characterized by the vector $\mathbf{p}$. In this case the force of attraction will be the same for all the variables, and then the distortion will depend only on the assumed dependence structure; if we work with the independence assumption, the arrival matrix will converge to the not distorted combinatoric distribution matrix which defines the CHC and the CHY approach for a sufficient large number of trials. In the base scenario, we decide to compute an arrival matrix of dimension $d = 2^{10}$ that corresponds to the number of combinatoric distributions of the number of defaults. Moreover we choose to work with a number of trials $N < d$. We analyze in the following also the case $N \geq d$.

In the following we model the dependence structure through a very common Archimedean copula, i.e., a Clayton copula function, and then we consider the

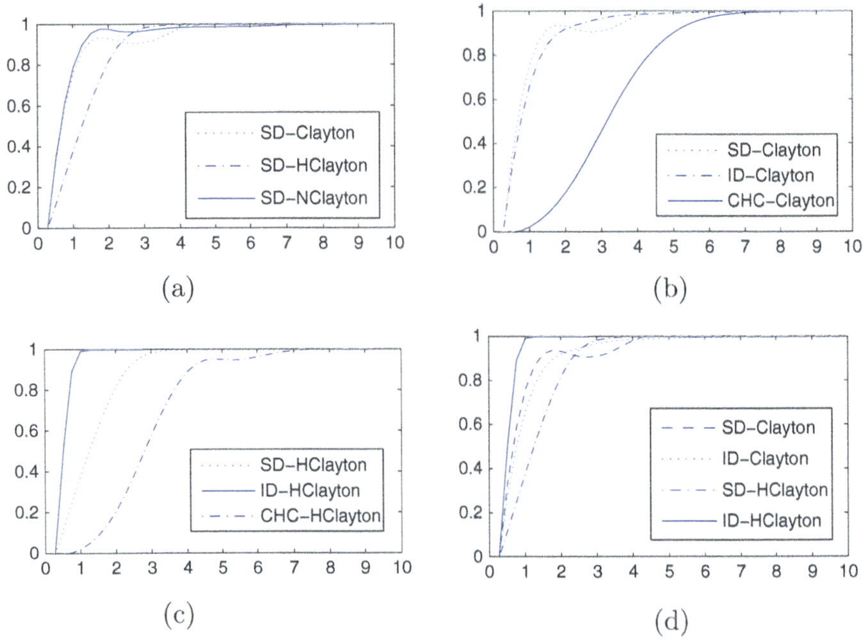

Fig. 5.3 Structural distortion: Clayton dependence. (**a**) Structural distortion on the loss cdf for several dependence structure: Clayton, HClayton, and NClayton. (**b**) Structural distortion on the loss cdf for Clayton structure vs ID-Clayton and CHC-Clayton. (**c**) Structural distortion on the loss cdf for HClayton structure vs ID-HClayton and CHC-HClayton. (**d**) Structural distortion on the loss cdf for Clayton and HClayton structure vs the corresponding ID-structure

different cases of standard Clayton, hierarchical, and nested Clayton (HClayton and NClayton, respectively) for the dependence parameters computed starting from the assumed Kendall τ^1 and $\boldsymbol{\tau}^{\mathbf{2}}$.

In Fig. 5.3a we plot the cumulative distribution functions[12] of the number of defaults computed with the DHC approach where we assume homogeneous marginal default probabilities. Here we see the impact of the dependence structure on the arrival policy that we call *structural distortion* (SD), for different dependence structures in the Clayton family. We can observe the dependence structures with at least two levels are in general more conservative than those having just one level since here the impact on the arrival matrix depends on a very small first-level τ^1, implying a feeble distortion.

In order to evaluate the impact of the dependence structure on the distortion of the loss cdf, we make a comparison of the loss distribution computed through the DHC approach with homogeneous marginals, with the case where also the arrivals

[12]The pictures are generated by interpolation and smoothing technic in order to represent the discrete distribution as continuous one.

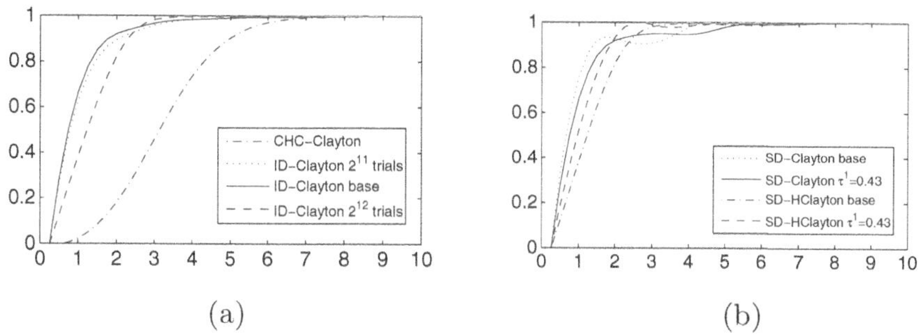

Fig. 5.4 Structural distortion: Clayton dependence. (**a**) ID-Clayton loss cdf for several number of trials vs CHC-Clayton. (**b**) Structural distortion for homogeneous Clayton and HClayton for different τ^1

are assumed to be independent of the dependence structure among variables (ID for short) and the CHC loss distribution. This last case does not generate any distortion since the arrival policy is not considered into the model. In Fig. 5.3b, c, we make the comparison considering the Clayton and the HClayton structure, respectively. The behavior of the loss cdfs shows that in general the result of a structural distortion is a more conservative curve. In picture Fig. 5.3d we evaluate the structural distortions for the Clayton and the HClayton structure. Starting from a given number of defaults, the impact of the distortion becomes greater in the hierarchical dependence since the second level of dependence is stronger than the first one which corresponds to the unique dependence level of the Clayton case (Fig. 5.4).

The ID case, that is, considered in the base scenario for an arrival matrix of dimension equal to the number of the combinatorial distribution of the number of defaults (i.e., 2^{10} in this example) converges to the CHC one, for a large number of trials, as we can observe in picture Fig. 5.5a for one-level Clayton structure.

Moreover the structural distortion depends on the dependence parameters. To assess this sensitivity, we consider a scenario characterized by an higher first-level Kendall $\tau^1 = 0.43$ to be compared with the base scenario. In Fig. 5.5b we plot the loss cdfs corresponding to the homogeneous one-level and hierarchical Clayton structure into the two scenarios. The behavior of the curves shows that the effect of a greater first-level Kendall τ^1 is generally results into a less conservative loss cdf for the one-level structure and into a more conservative one for the hierarchical case.

As pointed out before, the impact of the arrival policy on the loss distribution depends also on the marginal default probabilities which represent the force of attraction of the default event. In Fig. 5.6a we plot the loss cdf for a set of not homogeneous variables where the dependence structure does not impact the arrival policy. This analysis allows to disentangle the effect of the inhomogeneous marginals on the arrival policy; this is a partial distortion that we call *marginal distortion*. The different dependence structures considered here act only into the aggregation step since we do not consider here any *structural distortion*.

The qualitative behavior of the curves shows that the marginal distortion does not impact in a stronger way into the dependence structure with at least two levels than into the one-level ones since here the dependence structure enters only in the aggregation step but does not modify the arrival matrix. On the other hand, the marginal distortion depends on the degree of dispersion of the marginal default probabilities around the centroid.

To analyze the sensitivity of the marginal distortion with respect to the dispersion of marginals, we consider a maximum dispersion scenario where the marginals are concentrated around the extreme default probabilities that are 0.005 and 0.5 in our example but where the centroid is the same of the base scenario. If the centroid does not change, obviously the structure distortion will not be modified, while the marginal distortion will be. We compare the loss cdfs in picture Fig. 5.6b; we can observe the impact of the greater dispersion is more evident in the two-level dependence structure since at the second level, the dispersion interacts with stronger degree of dependence. The two curves relative to the maximum dispersion scenario, which are more conservative than the base scenario, are also very similar implying that the effect of the marginals' dispersion prevails over that related to the dependence structure acting here only in the aggregating phase.

The *total distortion*, that is, the sum of the structural and the marginal distortions, is evaluated in Fig. 5.6 for several dependence structure. The totally distorted loss cdf is generally more conservative into the at least two-level dependence structure showing that into this scenario the prevalent distortion is the structural one.

In Fig. 5.6b, c, d we compare the partially distorted loss cdf with the totally distorted and the not distorted ones, for several dependence structure. Here it is possible to evaluate the impact of the distortion and to disentangle the effect of the structural and the marginal impact. The one-level Clayton structure shows the two partial distortions, the structural and the marginal ones, interact generating a total distortion curve which is less conservative than the partially distorted cdfs. On the other hand, the nested and the hierarchical structures produce a total distorted cdf which is intermediate between the SD-cdf, the most conservative one, and the MD-cdf. The not distorted curve is always less conservative than the totally distorted ones but not than the partially distorted since the MD-cdf is generally less conservative for the two-level dependence structures.

Finally we compare the base scenario loss cdf in one-level and two-level structure with a more dispersed scenario having a first level Kendall $\tau^1 = 0.43$. Here the total distortion sums up the effect of a greater dispersion on the marginal distortion and a greater dependence parameter on the structural distortion. In Fig. 5.7 we see the resulting final effect which generally produces a more conservative loss cdfs than the base scenario.

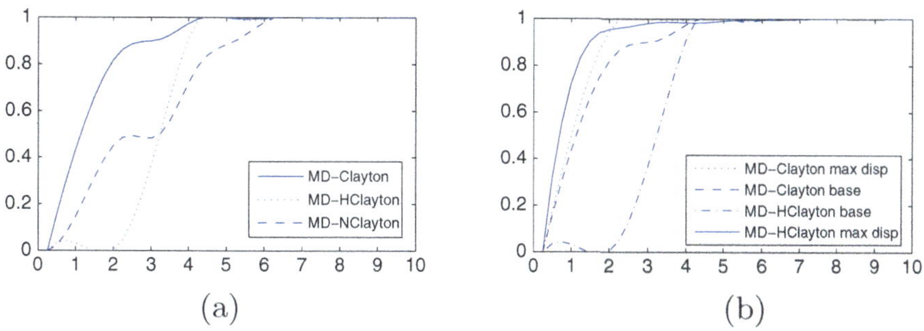

Fig. 5.5 Marginal distortion: Clayton dependence. (**a**) Marginal distortion on the loss cdf for Clayton, HClayton, and NClayton structure. (**b**) Marginal distortion on the loss cdf for Clayton and HClayton for different degrees of marginals' dispersion: base vs maximum dispersion scenario

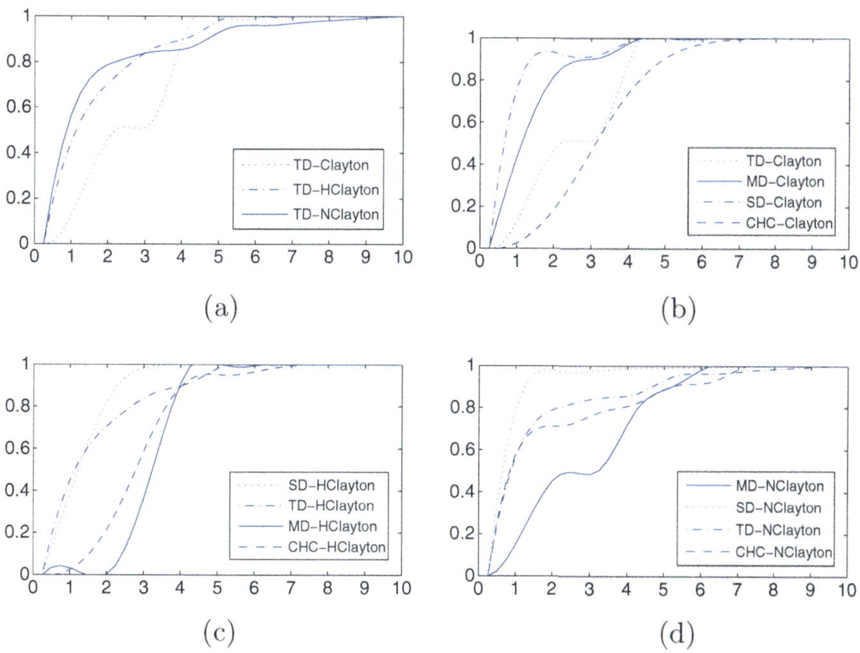

Fig. 5.6 Total and partial distortion: Clayton dependence. (**a**) Total distortion on the loss cdf for Clayton, HClayton, and NClayton structure. (**b**) Total and partial distortions on the loss cdf for Clayton structure vs the corresponding not distorted cdf. (**c**) Total and partial distortions on the loss cdf for HClayton structure vs the corresponding not distorted cdf. (**d**) Total and partial distortions on the loss cdf for NClayton structure vs the corresponding not distorted cdf

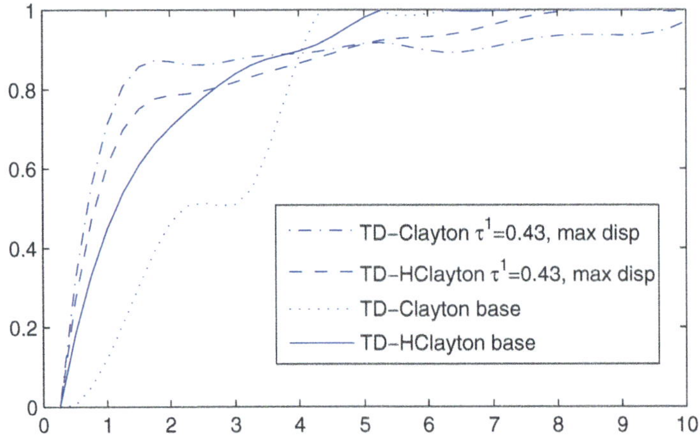

Fig. 5.7 Total distortions on the loss cdf for Clayton and HClayton structures: base scenarios vs the maximum dispersion one with $\tau^1 = 0.43$

5.2.2 A Pseudo-Spectral Analysis of the Arrival Matrices

As discussed above, the distortions of the loss cdf is induced by the arrival policy. We propose an analysis of the pseudo-spectrum, i.e., the set of the singular values of the arrival matrix, relative to different arrival policies, considering a number of trials N greater than the row's dimension of the arrival matrix d, i.e., $N = 2^{12}$. The idea behind this analysis is to show that the distortions of the arrival matrix act on the probability distribution of pseudo-spectra modifying the dispersion of the frequencies. A more dispersed case, corresponding to a more uncertain scenario, would generate a more conservative loss distribution. In this view a pseudo-spectral analysis of the random matrix could provide useful information to control and to manage the risk induced by the arrival policy of defaults. We consider the following cases:

- Independent copula, identically distributed marginals (I, id): this is the not distorted case.
- Independent copula, not identically distributed marginals (I, not id): this is a partially distorted case (marginal distortion).
- HClayton copula, identically distributed marginals (HClayton, id): this is a partially distorted case (structural distortion).
- HClayton copula, not identically distributed marginals (HClayton, not id): this is a totally distorted case (structural and marginal distortions).
- NClayton copula, identically distributed marginals (NClayton, id): this is a partially distorted case (structural distortion).
- NClayton copula, not identically distributed marginals (NClayton, not id): this is a totally distorted case (structural and marginal distortions).

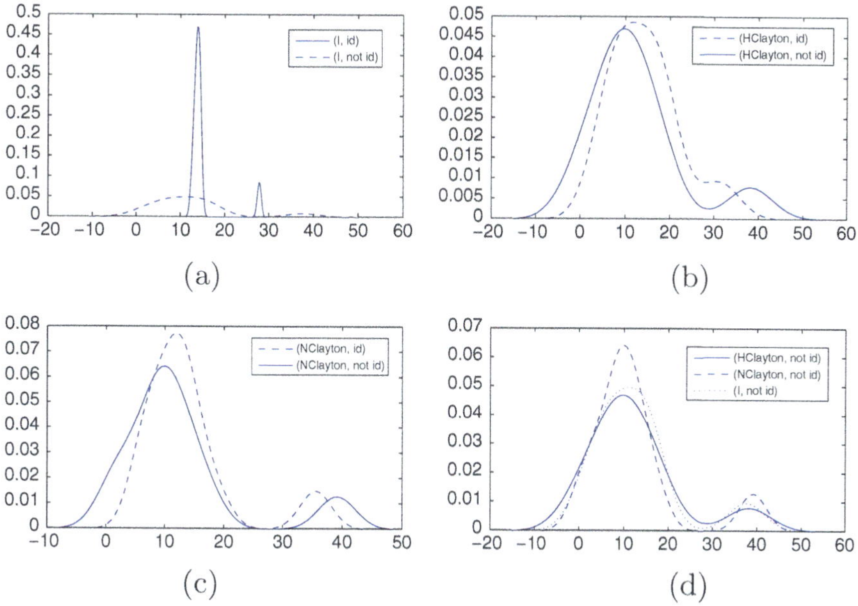

Fig. 5.8 Kernel smoothing density of the arrival matrix's pseudo-spectrum: Independence and Clayton dependence. (**a**) Kernel smoothing density of the arrival matrix's pseudo-spectrum: (I, id) versus (I, not id). (**b**) Kernel smoothing density of the arrival matrix's pseudo-spectrum: (HClayton, id) versus (HClayton, not id). (**c**) Kernel smoothing density of the arrival matrix's pseudo-spectrum: (NClayton, id) versus (NClayton, not id). (**d**) Kernel smoothing density of the arrival matrix's pseudo-spectrum: I, HClayton, and NClayton, with not id marginals

In Fig. 5.8a we can observe in the (I, id) case the singular values are all positives and very concentrated into two points,[13] while in the (I, not id) case, they are more dispersed, and the range of the pseudo-spectrum is wider. The same conclusion may be done also when we consider a not independent aggregation; the impact of the marginal distortion is a greater dispersion of the pseudo-spectrum (see Fig. 5.8b, c for the HClayton and the NClayton aggregation, respectively).

In Fig. 5.8d we compare the density of the pseudo-spectrum in case of independence, HClayton and NClayton aggregation, with different marginals; the curves are similar, but the most concentrated is the NClayton where the effect of the structural distortion partially compensates the impact of the marginals. In Fig. 5.9 we compare the pseudo-spectrum density accounting only for the structural distortion. We observe that the nested Clayton generates a distribution less dispersed than the hierarchical Clayton copula but more dispersed than the independence case where the probability mass is concentrated into two points. Moreover we can see the

[13]This evidence remembers the two-bulk configuration of the spectra of square and symmetric random matrices with iid entries (see Mehta 18).

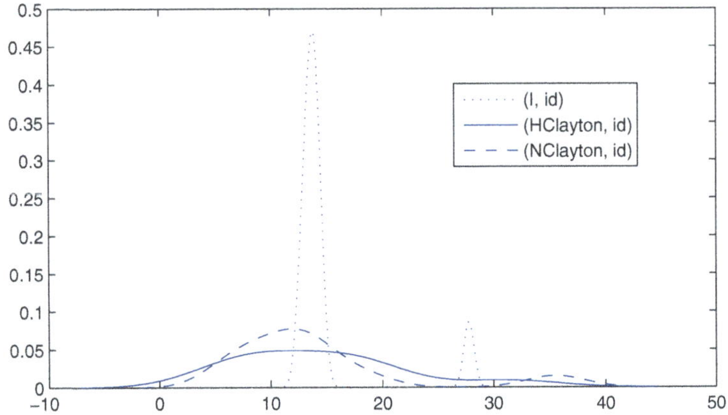

Fig. 5.9 Kernel smoothing density of the arrival matrix's pseudo-spectrum: I, HClayton, and NClayton, with id marginals

structural distortion generates a wider and more dispersed spectrum, and the effect of the structural distortion is not stronger than the impact of the marginals.

In Fig. 5.10a we compare the periodogram mean-square spectrum estimate in the not distorted case and in the marginally distorted one. It is evident that the marginal distortion acts, generating a flat periodogram less concentrated into some normalized frequencies, confirming the pseudo-spectra dispersion effect caused by the marginal distortion. As we see in Fig. 5.10b, c, the impact on the periodogram of the marginal distortion is weak if we consider a hierarchical or a nested Clayton dependence structure; the curves become a bit more flat. Moreover the total distortion induced by the hierarchical and nested Clayton copula is very similar (see Fig. 5.10d).

In Fig. 5.12a we evaluate the impact of the ratio of the number of trials, N and the rows' dimension of the arrival matrix, d[14] on the pseudo-spectrum distribution of the (I, not id) case. We see that if $N/d > 1$, the pseudo-spectrum distribution is more concentrated in two points, and then it is more similar to the iid case configuration, also if the two points are different (and define a wider range of the pseudo-spectrum) (Fig. 5.11).

In Fig. 5.12b we plot the distribution of the mean singular values for the (I,not id) case to analyze the effect of the marginal distortion for different ratio N/d. We observe that for a greater ratio N/d, the distribution becomes more regular and more concentrated into a singular value. Nonetheless it does not converge to the

[14]This ratio seems to be relevant in the results concerning the limiting spectra distribution of a random matrix.

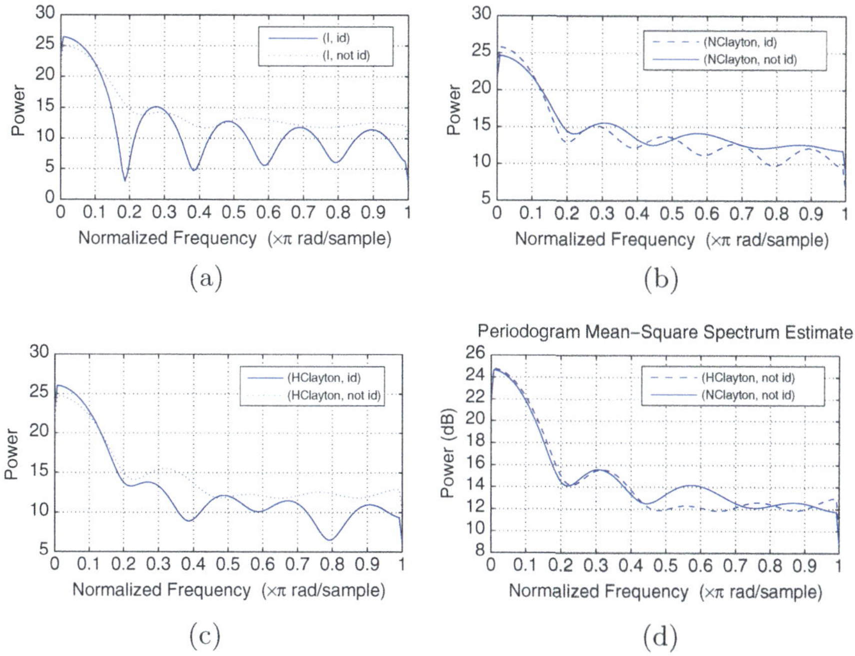

Fig. 5.10 Periodogram of the pseudo-spectrum. (**a**) Not distorted case versus the marginal distorted one. (**b**) NClayton structural distortion versus NClayton total distortion. (**c**) HClayton structural distortion versus HClayton total distortion. (**d**) HClayton versus NClayton total distortion

pseudo-Marcěnko Pasteur distribution, computed for the highest ratio.[15] Moreover we highlight that for $N/d = 1$ we have a marginal distorted distribution more similar to the pseudo-Marcěnko Pasteur distribution than the previous case with a greater value of N/d. In Fig. 5.12c, d we compare the mean singular values distributions for random matrices induced by a structural distortion and by a total distortion, respectively. As in the marginally distorted case, we have no convergence to the pseudo-Marcěnko Pasteur distribution when the ratio N/d increases (in fact in these cases, we see that the better performance is for $N/d = 0.1$).

Finally we study the dependence structure between the pseudo-spectrum induced by the different arrival matrices, fitting a copula and generating a random sample for the estimated dependence. Therefore we transform the random sample back to the original scale of the data and we analyze the pair correlations of the pseudo-spectra into the different arrival matrix assumption. In Fig. 5.12 we show the correlation between the simulated not distorted pseudo-spectrum and the partially and totally

[15]Marcěnko Pasteur distribution is the limiting distribution of the eigenvalues of a square symmetric random matrix with iid entries. We compute this limiting distribution into the pseudo-spectra range and for this reason we call it *pseudo-Marcěnko Pasteur distribution*.

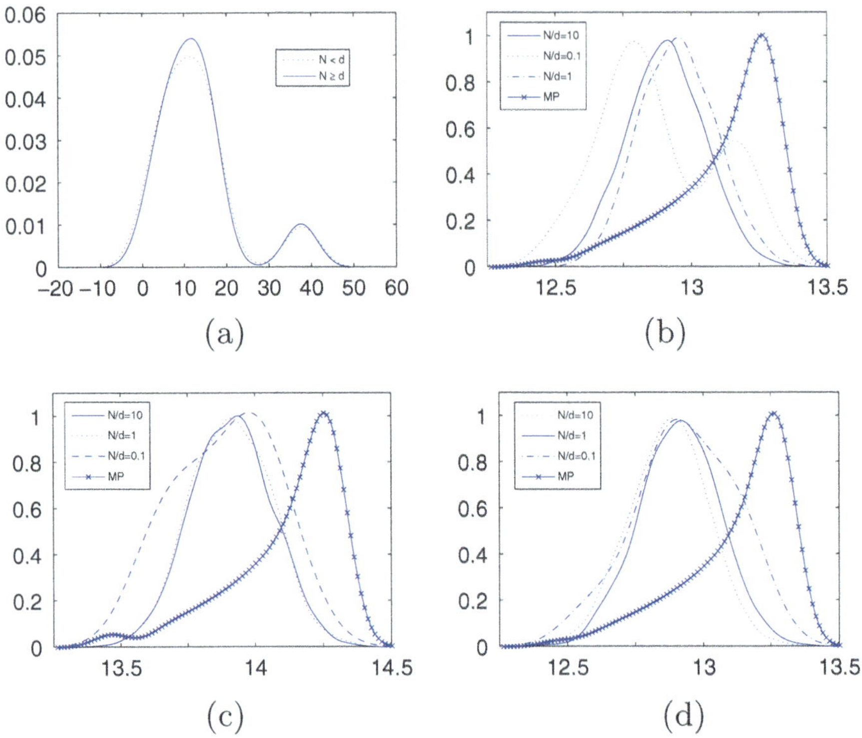

Fig. 5.11 Mean singular values distribution vs pseudo-Marcěnko Pasteur distribution. (**a**) Kernel smoothing density of the arrival matrix's pseudo-spectrum: (I, not id) for several number of trials. (**b**) Mean singular values distribution: (I, not id) for several number of trials vs pseudo-Marcěnko Pasteur distribution. (**c**) Mean singular values distribution: (HClayton, id) for several number of trials vs pseudo-Marcěnko Pasteur distribution. (**d**) Mean singular values distribution: (HClayton, not id) for several number of trials vs pseudo-Marcěnko Pasteur distribution

distorted ones. As expected we observe a relevant lower tail dependence, justified by the probability mass concentration on the right side of the not distorted distribution. We observe also that the marginal distortion (not id cases) induced a wider negative range of values of the pseudo-spectrum, while if only the structural distortion acts, this range is quite completely positive.[16]

Figure 5.13a, b explain the correlation between the partially distorted pseudo-spectra simulations, i.e., between the effects of the marginal and the structural distortions itself. We observe a strong right tail dependence also if we have some level of dependence on all the range of the pseudo-spectra particularly if we consider the hierarchical structure.

---

[16]We omit the comparison of the pseudo-spectra distribution in the not distorted case and the (NClayton, id) because their dependence structure is very similar to picture 5.12b.

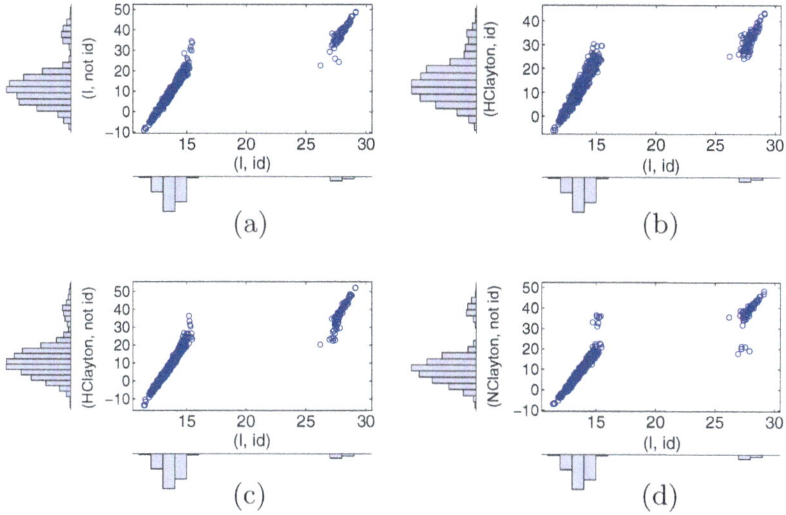

Fig. 5.12 Dependence between the pseudo-spectra distributions: not distorted case versus partially and totally distorted ones. (**a**) (I, id) versus (I, not id). (**b**) (I, id) versus (HClayton, id). (**c**) (I, id) versus (HClayton, not id). (**d**) (I, id) versus (NClayton, not id)

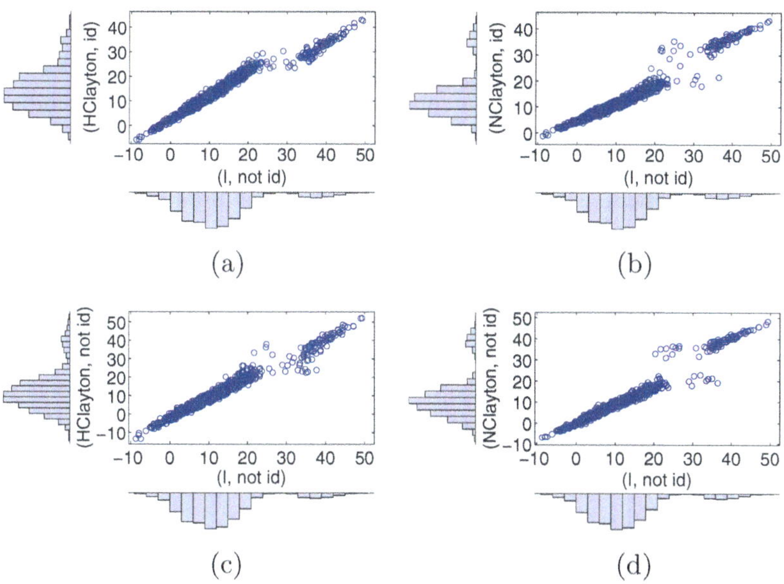

Fig. 5.13 Dependence between the pseudo-spectra distributions: marginal distorted case versus structural and totally distorted ones. (**a**) (I, not id) versus (HClayton, id). (**b**) (I, not id) versus (NClayton, id). (**c**) (I, not id) versus (HClayton, not id). (**d**) (I, not id) versus (NClayton, not id)

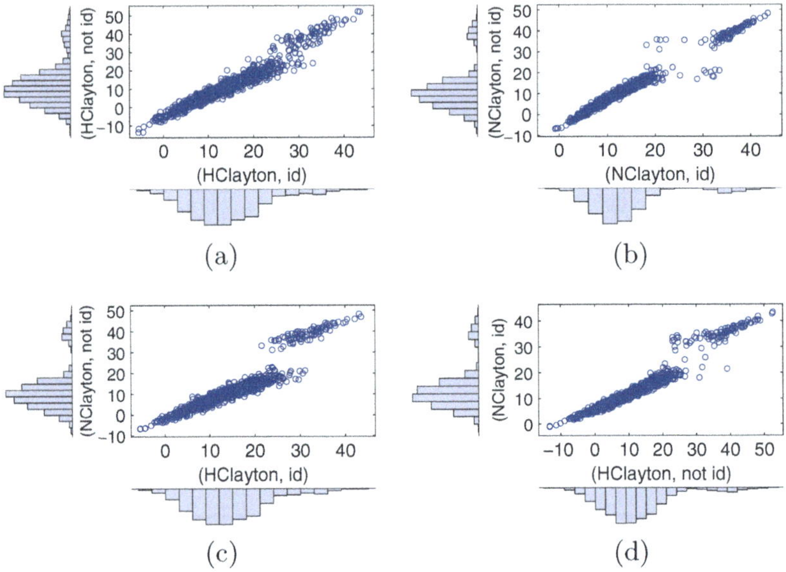

Fig. 5.14 Dependence between the pseudo-spectra distributions: structural distorted case versus totally distorted ones. (**a**) (HClayton, id) versus (HClayton, not id) (**b**) (NClayton, id) versus (NClayton, not id). (**c**) (HClayton, id) versus (NClayton, not id). (**d**) (HClayton, not id) versus (NClayton, id)

In Fig. 5.13c, d, we plot the pair correlations between the marginal distortion on the pseudo-spectrum and the total distortions induced by the hierarchical and nested Clayton dependence structure, respectively. The resulting evidence is very similar to the dependence between the marginal and the structural distortions on the pseudo-spectra. The structural distortion is then prevalent upon the marginal one.

Figure 5.14 shows the dependence structures between the partially distorted pseudo-spectra by the effect of a structural distortion and the totally distorted ones. The greatly diffused correlation on all the range of the pseudo-spectra is the one between the partially and totally distorted hierarchical structure.

Figure 5.15a compares the effects of two different structural distortions, while Fig. 5.15b shows the correlation between the total distortions induced by the previous structural distortions. These figures look very similar justifying the weakness of the marginal distortion impact included into a total distortion evaluation of the spectrum distribution (Fig. 5.16).

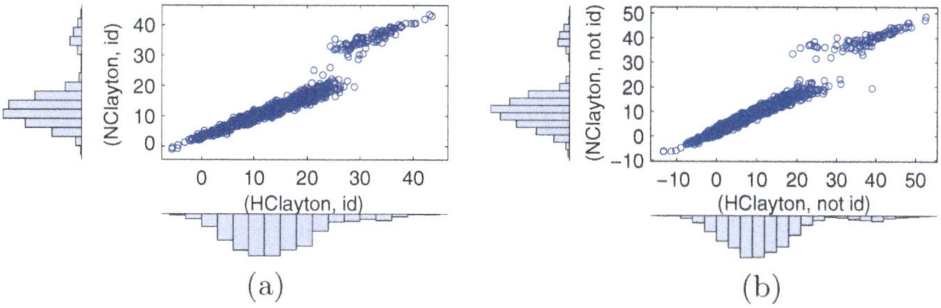

Fig. 5.15 Dependence between the pseudo-spectra distributions: structural distorted case versus structural distorted one and totally distorted case versus totally distorted one. (**a**) (HClayton, id) versus (NClayton, id). (**b**) (HClayton, not id) versus (NClayton, not id)

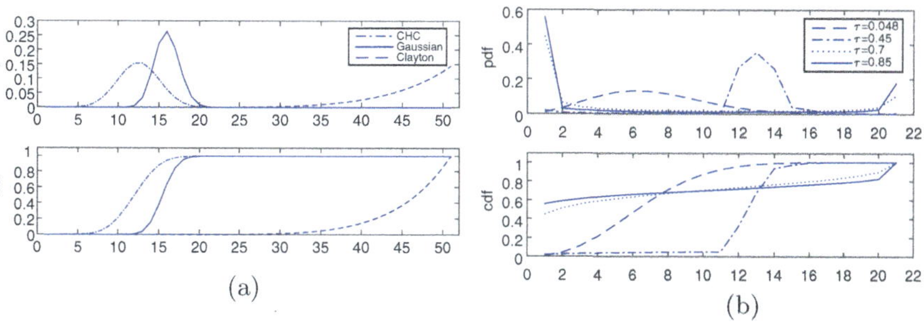

Fig. 5.16 Pdf and cdf of the defaults' number: CHC versus limiting models. (**a**) Pdf and cdf of the defaults' number. CHC versus PL Gaussian and Clayton models for a Kendall $\tau = 0.048$. (**b**) Pdf and cdf of the defaults' number through the CHC approach for several Kendall τ

5.3 High-Dimensional Problems: The Pure Limiting Models

The goal of this section is to focus on the problem to recover the distribution of a counting variable, linked to a high-dimensional problem which justifies the representation of its distribution through an approximated model. Obviously the kind of approximation detailing the *pure limiting models* to use depends from the field of application. Here we focus on credit risk applications.

Among the limiting models, we distinguish the pure ones (PL in the following) which are commonly used to compute the distribution of losses compliantly to Basel capital requirements. Here the key assumptions rely on having a homogeneous portfolio, and then an equal investment in each credit line, and a very large number of credits. These models are also called one-factor models because we assume that the default risks are correlated through a common factor or mixing variable which represents the systemic factor. This argument is supported by the structural Merton model. In this model the value of the debt of the ith firm depends on the firm's value which is formalized as

$$V_t^i = V_0^i \exp\left(\left(\mu^i - \frac{(\sigma^i)^2}{2}\right)t + \sigma^i\sqrt{t}\varepsilon^i\right),$$

for some $\mu^i, \sigma^i \in \mathbb{R}$ and where $\varepsilon^i = \frac{R_t^i - \left(\mu^i - \frac{(\sigma^i)^2}{2}\right)t}{\sigma^i\sqrt{t}}$ with $R_t^i = \log\left(\frac{V_t^i}{V_0^i}\right)$ is assumed to be $\mathcal{N}(0, 1)$. In Merton model the firm's values are correlated with a common market factor; if this correlation is supposed to be the same and equal to $\sqrt{\rho}$ for all firms, we can model the correlation as $\varepsilon^i = \sqrt{\rho}m + \sqrt{1-\rho}v^i$, where m represents the common factor and v^i the idiosyncratic one of the ith firm. Under different assumption on the joint distribution of m and v^i, we recover the following PL models.

We consider a large portfolio affected by one common factor and characterized by the following assumptions:

- We consider defaults up to a fixed-horizon time $T = 1$.
- There are I obligors having the same exposure size and the same loss in default. The number of obligors I is very large implying the relevant quantity for the portfolio risk is the fraction P_L of defaulted obligors in the portfolio.
- All obligors have the same unconditional default probability $p_i = p, \forall i$ until T.

1. Pure limiting Gaussian model. Regarding the common unconditional default probability, we recall that in the structural approach it is given by

$$p = \Phi\left(\frac{R_t^i - \left(\mu^i - \frac{(\sigma^i)^2}{2}\right)t}{\sigma^i\sqrt{t}}\right),$$

where $\Phi(.)$ denotes the cumulative standard normal distribution function. Then we can say the default occurs if

$$\varepsilon^i < \Phi^{-1}(p) \quad \text{or} \quad v^i < \frac{\Phi^{-1}(p) - \sqrt{\rho}m}{\sqrt{1-\rho}}.$$

We observe that conditional on the common factor m, the defaults are independent, and if we model the dependence with a Gaussian copula, the common conditional default probability is

$$\mathbb{P}(L^i = 1|m) = \mathbb{P}\left(v^i < \frac{\Phi^{-1}(p) - \sqrt{\rho}m}{\sqrt{1-\rho}}|m\right) = \Phi\left(\frac{\Phi^{-1}(p) - \sqrt{\rho}m}{\sqrt{1-\rho}}\right) = \theta,$$

where $L^i = 1$ if the i-th loan defaults and zero otherwise. For a given frequency of loss θ, we are able to calculate the value of the common factor m which will produce exactly this level of loss, i.e.,

$$m(\theta) = \frac{\Phi^{-1}(p) - \sqrt{1-\rho}\Phi^{-1}(\theta)}{\sqrt{\rho}}.$$

We observe that conditional on m, the variables L^i are independent and equally distributed with finite variance. This implies that $P_L = \frac{1}{N}\sum_{i=1}^{N} L^i$, which states for the portfolio's percentage loss conditional on m, converges by the law of large number to its expectation $\mathbb{P}(L^i = 1|m) = p(m)$ as $n \to +\infty$. Then finally the probability that the proportion of default P_L would be less than θ, i.e., the cdf of P_L, is

$$\mathbb{P}(P_L \leq \theta) = \mathbb{P}(p(m) \leq \theta) = \mathbb{P}(m > p^{-1}(\theta)) = \Phi(-m(\theta)),$$

where $m(\theta) = p^{-1}(\theta)$. If we stress the corresponding counting variable,[17] i.e., the number L of default occurred, we recover the default cdf, i.e.,

$$\mathbb{P}(L \leq q) = \mathbb{P}(m > m(\theta)),$$

where $L = NP_L$ and $q = N\theta$ where N is the number of credits. For this reason in Vasicek [22, 24], the distribution of the losses of a large loan portfolio[18] is described as follows

$$F(q) = \mathbb{P}(L \leq q) = \Phi\left(\frac{1}{\sqrt{\rho}}\left(\sqrt{1-\rho}\Phi^{-1}(q) - \Phi^{-1}(p)\right)\right), \tag{5.1}$$

where p is the default probability of any individual obligor, $\Phi(.)$ denotes the cumulative standard normal distribution function, and ρ is the asset value correlation between any two obligors induced by the correlation with one common market factor.

2. Pure limiting Archimedean model. An alternative to represent the limiting distribution of the loss consists in applying the Marshall-Olkin strategy. Marshall and Olkin [15, 16, 17] propose a method for the sampling of a random vector based upon the representation of the family of Archimedean copula function with Laplace transforms and mixtures of powers as described in Joe [13]. We apply the mixing-variable algorithm [17] to our problem. Its pivotal argument is that conditional on the realization of Y, i.e., the mixing variable, the random variables

[17] This is a limiting result which holds true for $n \to +\infty$, but it is accepted for *large* but finite n. The right dimension n which allows to classify it as *large* depends on the application at hand.

[18] There is no confusion here among default and loss cdf since they are the same but the support of the first is $\mathbb{N}$, while that of the second is $\mathbb{R}$.

$X_i := \phi^{[-1]}\left(-\frac{1}{Y}\log U_i\right)$, representing the marginals, where $U_i, i = 1, \ldots, I$ are independent of Y and i.i.d. uniformly distributed on $[0, 1]$, and where $\phi^{[-1]}(.)$ is the Y's Laplace transform, are independent. This conditional independence property is the fundamental point which drives the results in the credit risk model proposed in Schönbucher [19].

Proposition 5.3.1 (mixing-variable algorithm) *Let* $\phi^{[-1]} : \mathbb{R}^+ \rightarrow [0, 1]$ *and* $\phi : [0, 1] \rightarrow \mathbb{R}^+$ *be continuous, strictly decreasing functions. Follow the algorithm*

(a) *Consider* $U_1, \ldots, U_I$ *i.i.d. uniformly distributed on* $[0, 1]$.
(b) *Define the mixing variable* Y *such that:*

- G *and* g *are the* Y*'s distribution function and its density (if a density exists), respectively.*
- Y *is independent of* $U_1, \ldots, U_I$.
- $\phi^{[-1]}(.)$ *is the* Y*'s Laplace transform:*

$$\mathcal{L}_G(s) := \mathbb{E}\left(\exp(-sY)\right) = \int_0^{\infty} e^{-sy} dG(y) = \phi^{[-1]}(s). \tag{5.2}$$

(c) *Define*

$$X_i := \phi^{[-1]}\left(-\frac{1}{Y}\log U_i\right), \quad 1 \leq i \leq I, \tag{5.3}$$

representing the marginals of a set with cardinality I.

We note that

$$\mathbb{P}(X_i \leq x_i | Y) = \mathbb{P}(U_i \leq \exp(-\phi(x_i)Y)) = \exp(-\phi(x_i)Y).$$

Then the joint distribution function of X_i, $1 \leq i \leq I$ is

$$\begin{aligned}
\mathbb{P}(\mathbf{X} \leq \mathbf{x}) &= \mathbb{E}\left(\prod_{i=1}^{I} \exp(-\phi(x_i)Y)\right) = \mathbb{E}\left(\exp\left(-Y\sum_{i=1}^{I}\phi(x_i)\right)\right) \\
&= \phi^{[-1]}\left(\sum_{i=1}^{I}\phi(x_i)\right),
\end{aligned}$$

and the dependence structure of $X_i, i = 1, \ldots, I_j$ is represented by an Archimedean copula function with generator $\phi(.)$.

Given (5.2), it follows that the density of Y is the inverse Laplace transform of $\phi^{[-1]}$. It is important to observe that conditional on the realization of Y, the random variables X_i are independent.

In the limiting Archimedean approach, the loss distribution can be derived by conditioning on the mixing variable Y, representing the common risk factor which affects the loans. Conditionally on $Y = y$, the default probability of the ith obligor is

$$\begin{aligned} p_i(y) &:= \mathbb{P}\left(X_i \le p_i | Y = y\right) = \mathbb{P}\left(\phi^{[-1]}\left(-\frac{1}{y}\log U_i\right) \le p_i\right) \\ &= \mathbb{P}\left(U_i \le \exp(-y\phi(p_i))\right) = \exp\left(-y\phi(p_i)\right). \end{aligned}$$

By the large portfolio assumption and the law of large numbers, the fraction of defaults will almost surely be $p(y)$ in the limit, whenever the mixing variable Y has taken the value of y. Thus, the probability to have no more than q defaults in the portfolio is

$$\begin{aligned} F(q) &:= \mathbb{P}(L \le q) = \mathbb{P}(p(Y) \le q) = \mathbb{P}(\exp(-Y\phi(p)) \le q) \\ &= \mathbb{P}\left(Y \ge -\frac{\log q}{\phi(p)}\right) = 1 - G\left(-\frac{\log q}{\phi(p)}\right), \end{aligned} \tag{5.4}$$

where $G(.)$ is the Y's distribution function. Similarly the limiting loss density, if the mixing variable has a density g, is given by

$$f(q) = \frac{1}{q\phi(p)} g\left(-\frac{\log q}{\phi(p)}\right).$$

3. Panjer recursion approach. In *reliability literature* it is frequent to use common Poisson shock processes to model dependent event frequencies (see Barlow and Proschan 1 and Marshall and Olkin 15). In the same spirit, Lindskog and McNeil [14] examine a credit risk model where the frequencies of several kind of losses are related to a series of underlying and independent shock processes (in credit risk modeling, they might be a variety of economic events such as local or global recessions). When a shock occurs, this may cause losses of several different types; the common shock causes the numbers of losses of each type to be dependent. This approach generalizes the results for the multivariate Poisson process counting numbers of failures of different types which may be fatal or not necessarily fatal.[19] Anyway the model may be written in terms of an equivalent fatal shock model which facilitates the approximation of the aggregate loss frequency distribution using the *Panjer recursion* approach which also works

[19] We point out that if we count fatal events corresponding to default, we recover the default distribution, while if we count losses, we recover the loss distribution.

in higher dimensional problems. The total number of losses has the following stochastic representation:

$$N(t) \stackrel{d}{=} \sum_{i=1}^{\tilde{N}(t)} W_i,$$

where $W_1, \ldots, W_{\tilde{N}(t)} \stackrel{d}{=} W$, denoting the number of losses due to the i-th causing shock, are i.i.d. and independent of $\tilde{N}(t)$ which counts only the loss-causing shocks and which is a Poisson process with parameter $\tilde{\lambda}$. The pdf of the counting variable $N(t)$ can be easily calculated using *Panjer recursion*, i.e.,

$$\mathbb{P}(N(t) = r) = \sum_{i=1}^{r} \frac{\tilde{\lambda} t i}{r} \mathbb{P}(W = i)\mathbb{P}(N(t) = r - i), \quad r \geq 1,$$

where $\mathbb{P}(N(t) = 0) = \exp(-\tilde{\lambda} t)$.

Example 5.3.1 (PL models versus $1 - CHC$ *model: a credit risk experiment)* As we remarked before, our main aim has been to propose a procedure which may be useful to face high-dimensional problems as an alternative way to find the pdf of a counting variable. In order to assess the limiting models at work, we compare the pdf obtained through the CHC procedure with the limiting models discussed before. To this aim we consider the different models calibrated preserving the same Kendall τ. The limiting approach has been proposed mainly into the field of credit risk modeling. Hence we proposed an experiment in the credit risk field to assess the limiting models also for that concerns the scalability of the computing algorithm.
We consider here a large portfolio composed by credit exposed assets (i.e., a loan portfolio), and we assume that the event of default occurs if the losses are greater of a fixed level. The portfolio manager evaluates the marginal probabilities of default of each loan and, for structured finance purposes, is interested in recovering the pdf of the number of default, i.e., on the *default distribution*.

Given the large dimension of the portfolio, the loss distribution is not simply assessable. To reduce the complexity of the problem, we apply the $1 - CHC$ approach discussed before.
In this example we consider a portfolio composed by 50 loans grouped with K-mean clustering method, based on the Euclidean metric with tolerance[20] $\delta = 0.085$, into 3 groups with the following equivalent cardinalities:

$$ECardinalities = \begin{bmatrix} 35 & 10 & 5 \end{bmatrix}.$$

[20]The tolerance is defined as the maximum acceptable distance of the elements of a cluster from the centroid of the same cluster (see Bernardi and Romagnoli 2).

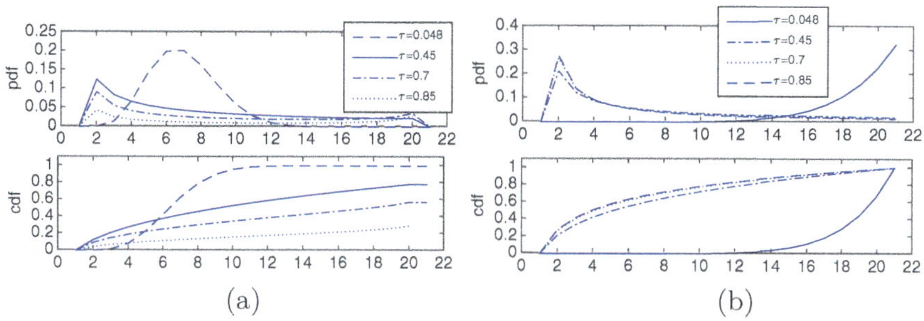

Fig. 5.17 Pdf and cdf of the defaults' number: CHC versus limiting models. (**a**) Pdf and cdf of the defaults' number with Gaussian PL model for several Kendall τ. (**b**) Pdf and cdf of the defaults' number with the Clayton PL model for several Kendall τ

These groups are assumed to be homogeneous with marginals equal to the centroids of the same group, which represents the rows of matrix **u** in Corollary 4.3.3:

$$Centroids = \left[0.25\ 0.28\ 0.35 \right].$$

The dependence parameter *within* and *between* the groups is $\alpha = 0.15$.

In Fig. 5.17a we plot the loss cdf and pdf computed through the CHC model and the Gaussian and the Clayton PL models for the same Kendall τ values and for a probability of default corresponding to the centroids' average.

The qualitative behavior of the pdf computed through the $1 - CHC$ model for $\tau = 0.048$ resembles the one recovered through the Gaussian PL models although the loss distribution function computed through the Clayton PL model appears always increasing without a single peak around the average default rate.[21] However, as pointed out in Schönbucher [20], the qualitative behavior of the default function is strictly linked to the level of dependence.

In Figs. 5.17b, 5.18a, b. we plot the loss pdfs and cdfs linked to the same reduced sample of 20 heterogeneous elements, computed through the CHC and the limiting models for several levels of Kendall τ. We observe that increasing the dependence's level, there is an impact on all the pdfs causing a shift toward the left due to a decrease in the average default rate.

Furthermore we compute the default distribution through the *Panjer recursion* where we assume to have only one type of loss-causing shock and where we calibrate the distribution of $\tilde{N}(t)$ with the joint probability of survival computed through a CHC. Figure 5.19a plots the loss distribution of the reduced sample with cardinality 20 which has been already considered before, computed through the *Panjer recursion* for several Kendall τ; an increase of the dependence's level

[21] In Clayton PL model, we recover a pdf with a single peak which is near to zero explaining the lower tail dependence of this copula function; here we cannot observe this peak due to approximation issues.

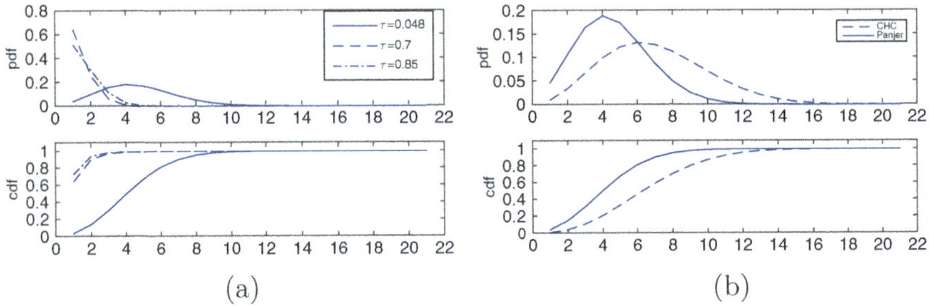

Fig. 5.18 Pdf and cdf of the defaults' number: CHC versus *Panjer recursion*. (**a**) Pdf and cdf of the defaults' number computed with *Panjer recursion* for several Kendall *tau*. (**b**) Pdf and cdf of the defaults' number. CHC versus *Panjer recursion* for a Kendall $\tau = 0.048$

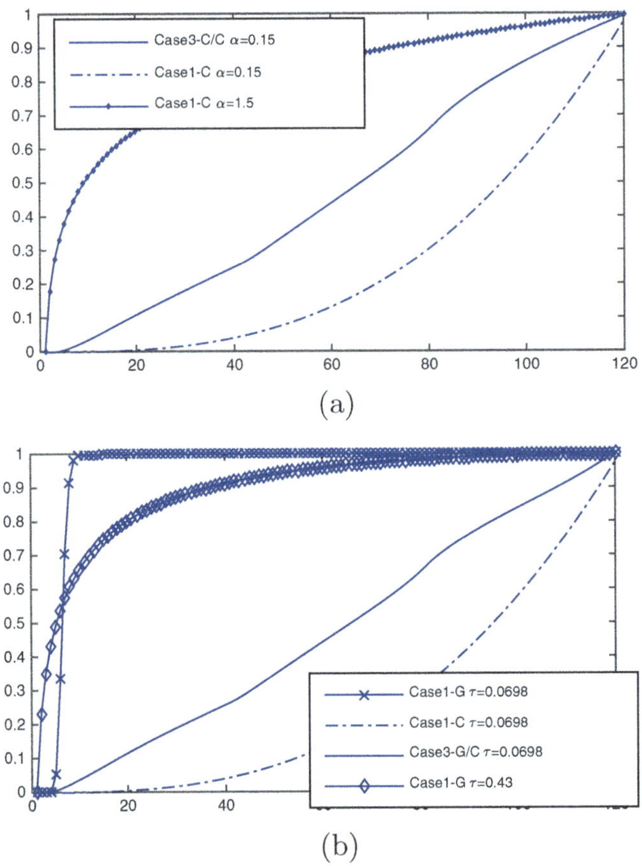

Fig. 5.19 Los distribution in the limiting cases: case 1 versus case 3. (**a**) Loss distributions computed in case 1 and case 3: Clayton limiting distribution for several dependence's levels vs Clayton limiting within and Clayton between groups. C/C stands for Clayton within/Clayton between. (**b**) Loss distributions computed in case 1 and case 3: Gaussian limiting distribution for several dependence's levels vs Gaussian limiting within and Clayton between groups. G/C in the legend stands for Gaussian within/Clayton between

impacts all the pdfs causing a shift toward the left, also for the *Panjer recursion* approach.

Figure 5.19b compares the default distribution computed through the *Panjer recursion* with the CHC approach for the same value of Kendall τ; we can observe that the behavior of the default distribution computed through the *Panjer recursion* is quite similar to the CHC approach for all the dependence levels considered in the experiments.

Finally we discuss the scalability of PL versus CHC default models. We consider the CHC model for the maximum dimension to which our techniques can be extended and discuss the scalability of the K-means CHC algorithm with respect to the number of groups and the total dimension of the problem. The MatLab™ algorithm was launched on a Macintosh HD, 2,8 GHz Intel Core i7 quad-core 16 GB 2133 MHz LPDDR3. Table 5.1 reports the computing time for the first 10 percent of the loss distribution corresponding to the reduced sample of dimension $n = 20$ for different numbers of groups and the CPU time-saver of the *Panjer* and PL approaches with respect to the CHC method.

We can observe that the number of groups is a critical variable; the CPU time increases exponentially with the number of group till 20 and for higher cardinality, problems of memory stack become unmanageable. Indeed we observe that *Panjer* and PL approaches are faster than CHC; the CPU time-saver corresponds to the price to have a better approximation as it is shown in Table 5.2. Here we compare the relative errors of the loss distributions's deciles computed through different methods with the exact distribution corresponding to our benchmark (i.e., the loss distribution computed through the single groups CHC model implemented by the CR algorithm detailed in Chap. 2) for a Kendall $\tau = 0.048$.

Table 5.1 CPU computation time in seconds and CPU time-saver with respect to the K-means CHC algorithm of the left queue (till 10 percent) of the loss distribution with *Panjer* and PL approaches. The dimension of the set is $n = 20$

N. of groups	Cpu time	STime Panjer	STime Gaussian	STime Clayton
2	0.53	0.55	0.13	0.12
4	20.02	20.07	21.06	10.56
6	131.52	134.52	134.15	104.73

Table 5.2 Percentage relative errors of the loss distributions' deciles computed with different models with respect to the CR loss distribution. Dimension of the set $n = 20$ and Kendall $\tau = 0.048$

Model	Relative error
CHC (3 groups)	35%
CHC (6 groups)	5.5%
Panjer	62%
PL Gaussian	86.5%
PL Clayton	97%

The CR loss distribution is computed considering that each element is characterized by its one default probability and then it corresponds to the *true probability distribution*. Unfortunately this model is unfeasible for high-dimensional problems that may be attacked by a reduction of its complexity. Here we compare the best performance of several models suggesting different solutions conformed to our simplification purpose, for a fixed level of dependence.

We specify that the reduced sample considered here is constructed assuring that the best performance of the PL models is obtained with a default probability p equal to the mean of the p_i, characterizing the heterogeneous sample. This implies that we fix the Kendall τ value and we obtain the best fit setting p as the mean of p_i. Then from τ, we can analytically recover the value of the dependence parameters, ρ and α. In a more complex example, a *good calibration* can improve the performance of the model; it could be possible to fix τ and calibrate p or also to fix p and calibrate τ, analytically linked to the dependence parameters. On the other hand, *Panjer recursion* is calibrated such to implicitly recover $\tilde{\lambda}$ from the joint probability of the no defaults scenario, computed through the CR model.

Finally regarding the CHC model, we recall the clusterization and the homogeneous assumption are compensated by a different cardinality of each set in order to preserve at the best the same dependence structure of the *true distribution*. This process allows to define the *equivalent* cardinality of each group ensuring the best calibration with respect to the CR loss distribution. In Table 5.2 we compare the performance of the CHC for a different number of groups, whose cardinality must be considered *equivalent* in the meaning specified in Chap. 2 (see Cifuentes and O'Connor 7 and Cifuentes et al. 8). We notice that the performance of the CHC method is better than other models. Clearly all the considered models loose a part of the information of the *true distribution*, as a price to reduce the complexity of the problem. As it is evident from Table 5.2, the loosing information, i.e., the percentage of information that is not explained by the CHC, decreases if the number of groups increases. To have a best fit, it is possible to choose a greater number of groups, considering the price of this best performance as shown in Table 5.1. An optimal criterion to choose the number of groups is to minimize the *confusions* of the set, as suggested in Wang et al. [25].

5.4 High-Dimensional Problems: The Limiting Clusterized Copulas

In this section we detail the limiting version of the clusterized copulas discussed in Chaps. 2 and 4. More precisely we'll introduce two new classes of copulas recovered as the limit of the CHC and the CHY, respectively. The first one called *Hierarchical limiting copula* (HLC for short) can be considered a limiting version of a CHC where we assume to have a proper hierarchy, i.e., more than one level of dependence. The idea is here to clusterize the underlying and then apply a multiple level CHC whose marginal cdfs are recovered through a PL model. On the other hand, the second one called *Hierarchical Hybrid copula* (HYC for short) is the limiting version of the CHY. Here only some classes are considered *large* and then formalized by a pure limiting model, while the others are evaluated by the usual combinatoric approach.

We'll describe these copulas in a credit risk application.

5.4.1 Hierarchical Limiting Model: A Credit Risk Application

This kind of copula has been suggested in credit risk field (see Bernardi and Romagnoli 3) to compute the loss distribution of a portfolio composed by correlated sub-portfolio with infinite granularity, which are differently affected by losses' events. Obviously if the CHC copulas don't consider the position inside the subgroups thanks to the homogeneous assumption, we could now have a greater simplification since we can skip the combinatoric calculus inside the sub-copulas because their marginal will corresponds to a limiting cdfs.

Usually in credit applications, all groups are considered enough large to have infinite granularity. Each group j satisfies the following conditions:

- It is composed by I_j obligors having the same exposure size and the same loss in default. The number of obligors I_j is very large implying that the relevant quantity for the portfolio risk is the fraction L^j of obligors in the jth sub-portfolio that suffer a loss.
- All obligors have the same unconditional loss probability $p_i^j = p^j, \forall i$ until a fixed-horizon time T.

The previous assumptions justify the application of a PL model to compute the jth loss cdf.

As seen before the PL Gaussian model suggested in Vasicek [22, 24] proposes the following loss cdf within the j-th large group

$$F_j^l(q) = \mathbb{P}(L^j \leq q) = \Phi\left(\frac{1}{\sqrt{\rho_j}}\left(\sqrt{1-\rho_j}\Phi^{-1}(q) - \Phi^{-1}(p^j)\right)\right), \tag{5.5}$$

where p^j is the loss probability of any individual obligor in the j-th class, $\Phi(.)$, denotes the cumulative standard normal distribution function and ρ_j is the asset value correlation between any two obligors of j-th class.

On the other hand, in the PL Archimedean approach, proposed in Schönbucher [19], the loss distribution of the j-th class can be derived by conditioning on the mixing variable Y^j representing the common risk factor affecting the j-th loans' group.

The probability to have no more than q losses[22] in the j-th portfolio is

$$F_j^l(q) := \mathbb{P}(L^j \leq q) = 1 - G_j\left(-\frac{\log q}{\phi_j(p^j)}\right), \tag{5.6}$$

where $G_j(.)$ is the Y^j's cdf and the limiting loss pdf of the j-th group is

$$f_j^l(q) = \frac{1}{q\phi_j(p^j)} g_j\left(-\frac{\log q}{\phi_j(p^j)}\right),$$

where g_j is the pdf of the mixing variable.

Being that the basket of loans is clusterized into n large sub-portfolios or classes affected by different risk factors, the basic idea is now to introduce a hierarchical risks' structure allowing us to take into account the non-exchangeability of the variables. The hierarchical limiting approach is implemented by two steps: a *within class* computing step and a *Between Classes* aggregation step.

The *Within Classes* Computing Step

We consider a portfolio clusterized into a finite number of large classes and compute the *within classes* loss distributions by a PL approach as the Gaussian or the Archimedean ones.

The *Between Classes* Aggregation Step

We consider n large sub-portfolios of homogeneous risky loans, i.e., affected by the same risk factor. Following a PL approach, we note the limiting loss distribution and the density function of class j, $j = 1, \ldots, n$ as $F_j(.)$ and $f_j(.)$, respectively. We are interested in determining the distribution of the whole loss defined as the sum of the classes' loss, $L = \sum_{j=1}^n L^j$.

Now the problem is to associate a counting variable to the copula function representing the dependence structure between the classes' loss distributions in order to distribute the losses through the classes.

Our aim is to consider a copula structure also for the *between classes* dependence such to recover a representation allowing us to take into account a hierarchical structure of risks. Via this hierarchical structure, we can extend the PL models to a non-exchangeable setting. It is well-known that the PL Gaussian and Archimedean models are exchangeable, i.e., the dependency between any two (or i) different risk factors does not depend on the question which risk factors were chosen. This is not a restriction for large homogeneous portfolio; however, it is not the case now, if we

---

22 We choose a properly defined unit of measure for the losses.

assume that the portfolio is clusterized into several groups representing the classes at different risk levels.

Giving the vector of the PL loss distribution within classes $\mathbf{F}(\mathbf{q})$ or the mixing variables' Laplace transforms vector for the multi-mixing model

$$\mathcal{L}_{G_1,\dots,G_n}\left(\sum_{i=1}^{I_1}\phi_1(x_i^1),\dots,\sum_{i=1}^{I_n}\phi_n(x_i^n)\right)=[A_1,\dots,A_n],$$

where $I_j, j = 1,\dots,n$ are the classes' cardinalities and $A_j = \phi_j^{[-1]}\left(\sum_{i=1}^{I_j}\phi_j(x_i^j)\right)$, we are interested in aggregating these functions with a dependence structure. By Sklar's theorem, since these functions are uniformly distributed, we can represent their dependence structure as a copula $HLC(F_1(q),\dots,F_n(q))$. This is a kind of hierarchical structure which characterizes the highest level of dependence.

For comparison purposes we introduce here two common assumptions that are frequently introduced in credit risk models. The first one which characterizes the one-factor models (see Gordy 10, 11 and Emmer and Tasche 9) is the following:

Definition 5.4.1 (Exchangeable between classes assumption) The exchangeable between classes assumption states that the classes are represented by the same loss probability $p^j = p, \forall j$.

On the other hand, a second assumption refers to the way the losses may distribute into the groups. It is the following:

Definition 5.4.2 (Equally affected policy) The equally affected assumption states that the probability that a loss attaches a group i is the same of group $j, \forall i, j$, i.e., the groups are equally affected by the losses' occurrence. This assumption implies that $L^i = L/n, i = 1,\dots,n$ as the size of each group gets to infinity.

We observe that an equally affected policy implies that we have total contagion between classes for that concerns the arrivals of the random event.

Example 5.4.1 In the HLC Archimedean model with equally affected assumption, the whole loss distribution function becomes

$$\begin{aligned}\mathbb{P}(L\le q)&=\mathbb{P}\left(Y^1\ge-\frac{\log q-\log n}{\phi_1(p^1)},\dots,Y^n\ge-\frac{\log q-\log n}{\phi_n(p^n)}\right)\\&=\overline{C}\left(G_1\left(-\frac{\log q-\log n}{\phi_1(p^1)}\right),\dots,G_n\left(-\frac{\log q-\log n}{\phi_n(p^n)}\right)\right),\end{aligned}$$

where $G_j(.)$ is the j-th mixing variable's distribution function for $j = 1,\dots,n$ and $\overline{C}$ is the survival of a copula function C representing the dependence structure between the losses for the highest hierarchy level of dependence.

To disentangle the implications of the different assumptions, we consider three settings explaining different impact on the loss distribution. We consider the following cases:

1. : Large number of large classes
2. : Finite number of large classes with equally affected assumption
3. : Finite number of large classes

Case 1

We observe that in this case the equally affected assumption is naturally included. If we accept also the exchangeability assumption, we can recover a *between* structure similar to the *within* one. This is the case of infinite granularity and homogeneous portfolio which justifies the application of a PL model to recover the whole loss distribution.

Example 5.4.2 In the *within* PL Archimedean model, we can imagine to define a mixing variable Y^W allowing to recover the conditional independence also between groups, i.e., such that

$$p^j(Y^W) = \mathbb{P}(A_j \leq a_j | Y^W) = \exp(-\phi_W(a_j)Y^W),$$

where $A_j := \phi_W^{[-1]}\left(-\frac{1}{Y^W}\log U_j\right)$ and $U_1, \ldots, U_n$ are i.i.d. uniformly distributed on $[0, 1]$ and independent of Y^W. Then we have

$$\begin{aligned}
\mathbb{P}(\mathbf{A} \leq \mathbf{a}) &= \mathbb{E}\left(\prod_{j=1}^{n} \mathbf{P}(A_j \leq a_j | Y^W)\right) = \mathbb{E}\left(\prod_{j=1}^{n} \exp(-\phi_W(a_j)Y^W)\right) \\
&= \mathbb{E}\left(\exp(-Y^W \sum_{j=1}^{n} \phi_W(a_j))\right) = \mathcal{L}_{G_W}\left(\sum_{j=1}^{n} \phi_W(a_j)\right) \\
&= \phi_W^{[-1]}\left(\sum_{j=1}^{n} \phi_W(a_j)\right),
\end{aligned}$$

where we observe that the variables A_j, $j = 1, .., n$ are linked by an Archimedean copula function with generator ϕ_W.

In case 1, if $a_j = a, \forall j = 1, \ldots, n$ and $p^j(Y^W) = p, \forall j = 1, \ldots, n$, i.e., under the exchangeable assumptions, we recover the following whole loss distribution and density functions:

$$F^W(q) = 1 - G_W\left(-\frac{\log q}{\phi_W(a)}\right)$$

$$f^W(q) = \frac{1}{q\phi_W(a)} g_W\left(-\frac{\log q}{\phi_W(a)}\right),$$

where g_W is the density of the whole mixing-variable, if it exists.

Case 2

We observe that in this case the whole loss distribution corresponds to the loss distribution defined by the groups' centroids, i.e., by the elements representing each group with mixing variable Y^j and loss probability $p^j(Y^j) = e^{-\phi_j(p_j)Y^j}$, $j = 1, \ldots, n$:

$$F^W(q) = HLC(F^1(q/n), \ldots, F^n(q/n)),$$

where HLC is the copula function representing the *between* groups dependence structure and $F^j(q/n) = 1 - G_j\left(-\frac{\log q - \log n}{\phi_j(p_j)}\right)$. As in the Example 5.4.1, here we assume that, if the j-th group registers q/n defaults, the same happens for other groups.

Finally we observe that the dimension of the problem is considerably reduced given the *equally affected assumption*. This setting corresponds to the assumption of a sectorial division into infinite granular portfolios equally affected by the defaults' events.

Case 3

This setting corresponds to the assumption of an infinite granularity of the correlated sub-portfolio which is differently affected by defaults' events. The affected policy is not commonly considered into the sector concentration risk.

In this case we propose the application of the HLC model based on a counting variable that counts unit of percentage losses.

Our aim is to recover the cumulated probability distribution of the counting variable linked to the n-dimensional copula function HLC. To do it we observe that all the c.c.ds of l ones into n groups correspond, through its probability function, to a number in $[0, 1]$. We enumerate the c.c.ds by an index and observe that the j-th c.c.d. gives the vector $\mathbf{p}_{l,n}(j)$ and, finally through a copula function, it corresponds to a joint probability. This joint probability refers to the scenario to have a loss less than or equal to l percentage units distributed into the groups in the way explained by the j-th c.c.d.. We observe that the limiting approach (case 1) does not allow to specialize the kind of distribution considered, since the loans are considered exchangeable and equally affected, i.e., they are not clusterized into groups. The mean of the joint probabilities computed for all the c.c.ds for the same number of ones, i.e., for the same l ones into n groups, represents the (mean) cumulated probability to count l ones, or better to have l events whose probabilities are the marginal limiting probabilities of the groups, if we assume the different ways to distribute the events are equally probable. Anyway we may consider an expectation corresponding to the assumption related to the affecting policy of the events.

Proposition 5.4.1 (Expected cdf of the counting variable linked to a HLC function) *The cumulative probability distribution of the counting variable linked to the n-dimensional copula HLC, with n large and homogeneous groups of dimensions* $m_s, s = 1, .., n$, *respectively, such that* $\sum_{s=1}^{n} m_s = N$ *is the function* $F_r : [0, N] \to [0, 1]$ *such that*

$$F_r(i) = \mathbb{E}\left(HLC(\mathbf{F}(\mathbf{p}_{l,n}(j))), j = 1, \ldots, \hat{D}^c(i, n)\right),$$

where $\mathbb{E}$ *stands for a mean which is the average if the distribution is equally probable,*[23] $\mathbf{p}_{l,n}(j)$ *is a n-vector counting the ones corresponding to the j-th c.c.d. of l ones into n groups of dimensions* $m_s, s = 1, \ldots, n$, $\mathbf{F}(\mathbf{p}_{l,n}(j))$ *is the vector of limiting margins corresponding to the j-th c.c.d. of l ones into n groups, and* $\hat{D}^c(i, n)$ *counts the c.c.ds of i ones into n groups.*

We observe that when we have a hierarchical structure to represent the dependence among the random variables, to properly consider the distributions of the counting event, it would be necessary to distribute the event considering the affecting policy induced by the hierarchy. On the other hand, here we distributed uniformly the event and computed the mean loss distribution because our aim is to deal with the issues about the high complexity of the problem. Nevertheless we have already formalized in Sects. 5.1 and 5.2 the distribution policy introducing an arrival matrix corresponding to the distribution of the counting variable cdf induced by the hierarchy and the marginals.

Example 5.4.3 (Some experiments: PL versus HLC) In this example we consider a large portfolio composed by credit exposed assets, clusterized into 3 classes each one composed by 40 loans[24] grouped by a clustering method, based on the Euclidean metric with a little tolerance, for example, $\delta = 0.09$. These groups are

[23] If the different ways to distribute the events are equally probable, we must consider the average of copula values. Otherwise we can weigh the values coherently with our confidence assumptions.

[24] The cardinality of the groups justifies the large portfolio assumption since it is greater than the threshold of 20 (see Schönbucher 19).

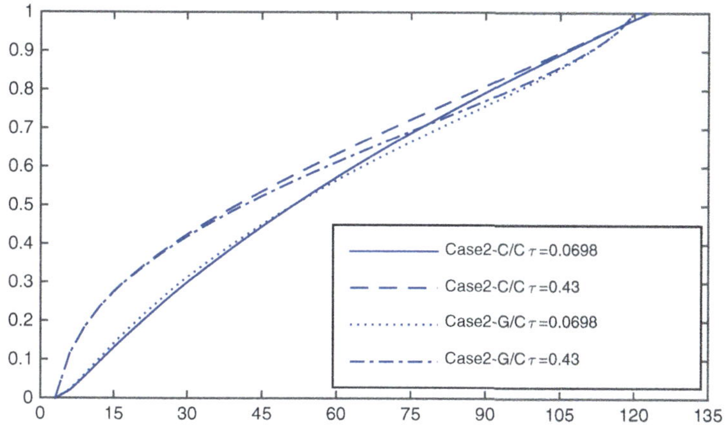

Fig. 5.20 Loss distributions computed in case 2: Clayton and Gaussian limiting distribution for several dependence's levels between groups. C/C in the legend stands for Clayton within/Clayton between while G/C stands for Gaussian within/Clayton between

assumed to be homogeneous with marginals equal to the centroids of the same group, which represent the vector of the default's probabilities:

$$Centroids = \begin{bmatrix} 0.028 \; 0.075 \; 0.047 \end{bmatrix}$$

The dependence parameters within the groups correspond to the following levels of Kendall τ: 0.6, 0.43, and 0.8 while the parameter between the groups corresponds to a Kendall $\tau = 0.048$. For example, if we assume to have a Clayton dependence structure within and between the groups, we would consider the following dependence parameters:

$$\alpha_{\text{within}} = \begin{bmatrix} 2.7 \; 1.5 \; 7.5 \end{bmatrix} \quad \alpha_{\text{between}} = 0.15$$

We compare the loss distribution computed following the approaches identified before as case 1,2, and 3. In case 1 we assume a probability of default equal to the mean of the centroids (we include also the *exchangeability assumption*) and a dependence level corresponding to the between groups dependence parameter.[25] In case 3 we assume to aggregate the limiting loss distributions within the groups through a copula function representing the highest level of the risks' hierarchy. This copula is linked to a variable which counts the percentage units of losses distributed between the groups.

In Fig. 5.20a, b we compare the loss distributions for the whole limiting case (case 1) for several dependence's levels, with the limiting case within groups/Clayton

[25]We express the dependence level in terms of Kendal's tau to compare directly the Clayton and the Gaussian case.

aggregation between groups (case 3), respectively, for Clayton and Gaussian limiting models. We can observe that case 1 with Clayton limiting model is more similar to case 3 for a dependence's level equal to the highest dependence parameter, i.e., the parameter of the copula function which aggregates the loss distributions within the groups. The whole Gaussian limiting model presents more similarities to case 3 for an higher dependence's level corresponding to $\tau = 0.43$; anyway the Clayton limiting model works better than the Gaussian one, especially for $\tau = 0.0698$. We can observe that case 1 Clayton loss distribution moves up if the dependence parameter increases, i.e., the mean of the distribution decreases, showing the property of lower tail dependence of Clayton copula. On the other hand, case 1 Gaussian exhibits the opposite effect since this distribution does not possess any tail dependence property (Fig. 5.21).

In Fig. 5.20 we compare the loss distributions computed through the Clayton and Gaussian limiting models with Clayton aggregation between groups (case 2) for several dependence levels. We can observe the distributions are very similar for the same dependence level and the order reflects the tail dependence property of the considered distributions (Clayton vs Gaussian). Moreover in case 2 an increment of the dependence level determines an upward movement of either the loss distribution, i.e., a lower mean level, showing the dominant effect is the one related to the lower tail dependence of the common Clayton copula (between groups aggregation).

In Fig. 5.22a, b we plot the loss distributions in case 2 vs the loss distributions in case 1 and 3, for Clayton and Gaussian limiting models, respectively, and for several dependence levels. We can observe that in case 2 the probability mass is more concentrated on the lower tail for higher dependence levels while in case 1 the resulting effect is related to the property of the considered distribution.

To better understand the difference between the loss distributions computed in the considered cases and evaluate the impact of the limiting assumptions, we make a comparison for several levels of percentage losses. In Fig. 5.23a, b we can appreciate the errors generated by the simplifying assumptions; the difference between the probability corresponding to case 1 and to case 3 is a *clusterization error*, including the *exchangeable error*, which is related to the *sector aggregation risk*, and the *equally affecting policy error*, while the difference between the probability corresponding to case 2 and to case 3 is only due to the *equally affected policy error*. The impact of the errors is greater for lower level of percentage losses and generally the *clusterization error* is greater in the Gaussian limiting model than in the Clayton one. On the other hand, the *equally affected policy error* appears lower in the Gaussian case for $\tau = 0.0698$, while it is comparable with the Clayton limiting model for higher level of dependence. We note also that the errors are evaluated considering as benchmark the case 3 with Clayton aggregation copula between groups, whose loss distribution clearly exhibits lower tail dependence. This observation explains the reason why Gaussian limiting model is affected by a greater *clusterization error*. To understand the result concerning the *equally affected policy error*, we must observe that the comparison depends on the considered dependence level since now all cases 2 reflect the lower tail property of the benchmark (i.e., case 3).

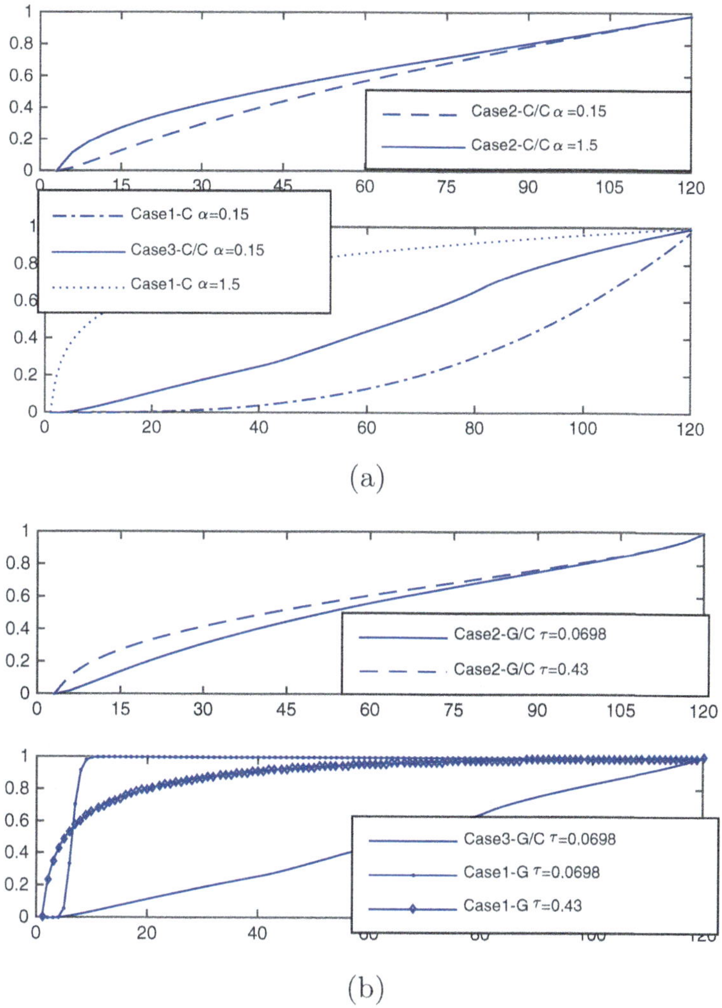

Fig. 5.21 Loss distribution in the limiting cases: case 1 versus case 3. (**a**) Loss distributions computed in case 2 vs case 1 and 3 for Clayton aggregation between groups: Clayton limiting distribution for several dependence's levels between groups. C/C in the legend stands for Clayton within/Clayton between. (**b**) Loss distributions computed in case 2 vs case 1 and 3 for Clayton aggregation between groups: Gaussian limiting distribution for several dependence's levels between groups. G/C in the legend stands for Gaussian within/Clayton between

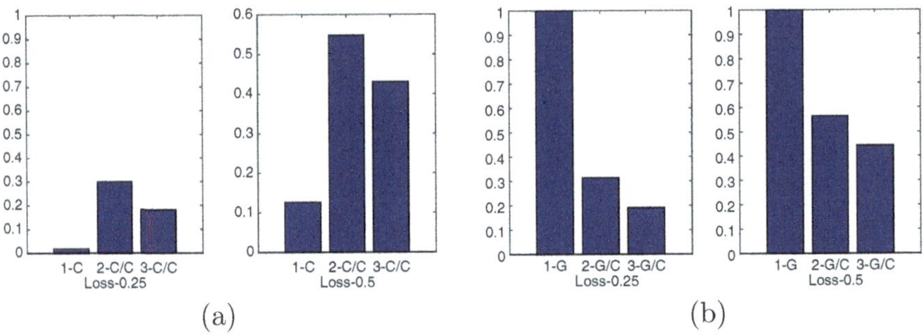

Fig. 5.22 Loss distribution in the limiting cases in cases 1, 2, and 3: aggregation between groups $\tau = 0.0698$. (**a**) Loss probabilities in case 1, 2, and 3 for Clayton limiting model within and Clayton aggregation between groups $\tau = 0.0698$: several percentages of losses. C/C stands for Clayton within/Clayton between. (**b**) Loss probabilities in case 1, 2, and 3 for Gaussian limiting model within and Clayton aggregation between groups $\tau = 0.0698$: several percentages of losses. G/C stands for Gaussian within/Clayton between

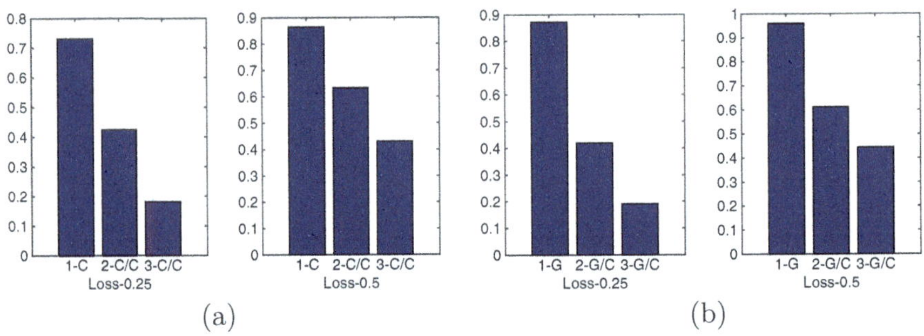

Fig. 5.23 Loss distribution in the limiting cases in cases 1, 2, and 3: aggregation between groups $\tau = 0.43$. (**a**) Loss probabilities in case 1, 2, and 3 for Clayton limiting model within and Clayton aggregation between groups $\tau = 0.43$: several percentages of losses. C/C stands for Clayton within/Clayton between. (**b**) Loss probabilities in case 1, 2, and 3 for Gaussian limiting model within and Clayton aggregation between groups $\tau = 0.43$: several percentages of losses. G/C stands for Gaussian within/Clayton between

5.4.2 Hierarchical Hybrid Copulas: A Credit Risk Application

Bernardi and Romagnoli [5] introduced the class of hierarchical hybrid copulas (HYC in the following) that corresponds to the HLC where the granularity is corrected for small groups.

Here we consider a credit risk model where we assume that the basket of loans is clustered into a finite number n of sub-portfolios or classes affected by different risk factors. The basic idea is to introduce a hierarchical risks' structure allowing to take into account for the non-exchangeability of the variables and for the grains of such classes. HYC model is implemented by the following steps: a preparing phase concerning the clustering and the correction for the homogeneous assumption as detailed in Chap. 2, a *within class* computing step and a *between classes* aggregation step.

- The *within classes* computing step. We consider a portfolio clustered into a finite number of classes and compute the *within classes* loss distributions. If the class is *large*[26] we compute the corresponding loss distribution by a limiting approach like as the Gaussian or the Archimedean ones. On the other hand, if the class is *small*, we recover the loss pdf through the 1-CHC approach.
  - *Large* classes. We assume that the j-th *large* class justifies the application of a PL model to compute the jth loss cdf referring to the r.v. L^j.

    Moreover we recall that it is possible to choose a PL Gaussian model as suggested in Vasicek [21, 23], or a PL Archimedean model, as proposed in Schönbucher [20]. We remind to the previous Sects. 5.3 and 5.4 for details about the formalization of such limiting loss distributions.
  - *Small* classes. If the class is *small*, we compute the relative loss pdf through the CHC approach. We review in the following the essentials about this approach which allows to extract the loss pdf from the dependence structure of the elements of a *small* class. This dependence structure is modeled through a $1 - CHC$, and then we provide a second level of clusterization inside the *small* groups. We assume that the jth *small* class is composed by n_j elements clusterized into k_j homogeneous groups of equivalent dimension $m_s^j, s = 1, .., k_j$, respectively, such that $\sum_{s=1}^{k_j} m_s^j = n_j$. The most important ingredient to extract the loss pdf from a CHC is the volume of this kind of copula.

    As pointed out before, we are interested on an event which may involve a basket of variables, composing a *small* class, and on defining a counting variable on this event that we call *second level counting variable*.[27] We observe that the *second level counting variable* counts the number of unit percentage losses inside the jth *small* class and then corresponds to the variable L^j we have already introduced and discussed concerning the *large* classes.

Definition 5.4.3 (Second level counting variable linked to CHC_j) The counting variable linked to the $CHC_j(\mathbf{c}(\mathbf{p}_{l,k_j}(z)))$ is given by

$$r(z) = \sharp(\mathbf{p}_{l,k_j}(z)), \forall j$$

which counts the number of elements equal to one in the matrix $\mathbf{p}_{l,k_j}(z)$ that distributes l units of loss in k_j groups, for the z-th c.c.d.. Clearly we have $r(z) = l, \forall z$.

[26] We say that the class is *large* if its cardinality is greater than or equal to a fixed bound.

[27] This is a *second level counting variable* since it is linked to a second level dependence structure or a dependence structure within a class.

Our aim is to extract the loss pdf of the jth *small* class from the CHC_j representing the dependence structure of the jth class. Hence we reproduce the usual approach also at this second level of dependence.

Corollary 5.4.2 (Loss pdf linked to the CHC_j) *The loss pdf linked to the CHC_j with centroids $\mathbf{u} \in [0,1]^{k_j}$ is the function $f_j^s : [0,n] \to [0,1]$ such that*

$$f_j^s(i) = \sum_{z=1}^{\hat{D}^c(i,k_j)} V_{CHC_j}(\mathbf{S}^z(\mathbf{p}_{i,k_j}(z))),$$

where $\mathbf{p}_{l,k_j}(z)$ is a $k_j \times \max(m_s^j, s = 1, \ldots, k_j)$ matrix of zeros and ones corresponding to the zth c.c.d. of l ones into k_j groups of dimensions m_s^j, $s = 1, \ldots, k_j$ and finally accounting for the permutation of the elements into the groups and where $V_{CHC_j}(\mathbf{S}^z)$ is the volume of the CHC_j computed for the box $\mathbf{S}^z = [\mathbf{u}^z, \mathbf{v}^z] \in \mathbb{R}^a \times \mathbb{R}^a$ with $\max(a) = 2k_j$, determined for the zth c.c.d. of i ones into k_j groups following the rule:

- *If $p_{s,w} = 0$, $u_{s,w}^z = u_s$ and $v_{s,w}^z = 1$ for $s = 1, \ldots, k_j$ and $w = 1, \ldots, m_s^j$ and if $p_{s,w} = 1$, $u_{s,w}^z = 0$ and $v_{s,w}^z = u_s$ for $s = 1, \ldots, k_j$ and $w = 1, \ldots, m_s^j$.*
- *Since the values of the coordinates corresponding to each $p_{s,w}$ may be only two, we correspond to each coordinate a two-dimensional vector such that*

$$u_{s,w}^z, \forall w \to \mathbf{u}_s^z = [u_{s,1}^z, u_{s,2}^z] \quad v_{s,w}^z, \forall w \to \mathbf{v}_s^z = [v_{s,1}^z, v_{s,2}^z],$$

where $(u_{s,1}^z, v_{s,1}^z)$ represents the coordinates of the not empty 0-elements set,[28] *and $(u_{s,2}^z, v_{s,2}^z)$ the ones of the not empty one-element set in the matrix $\mathbf{p}_{i,k_j}(z)$.*
- *Finally we define the box through the coordinates' vectors $[\mathbf{u}_s^z, \mathbf{v}_s^z]$ as*

$$\mathbf{u}^z = row(\mathbf{u}_s^z) \quad \mathbf{v}^z = row(\mathbf{v}_s^z),$$

where the function $row : \mathbb{R}^k \times \mathbb{R}^m \to \mathbb{R}^{k \times m}$ is defined as

$$row\begin{pmatrix} x_1^1 & \cdots & x_1^m \\ \vdots & \vdots & \vdots \\ x_k^1 & \cdots & x_k^m \end{pmatrix} = \left[x_1^1 \cdots x_1^m \cdots x_k^1 \cdots x_k^m\right].$$

[28] In the case of empty set, we delete the correspondent vector element. For example, if the set of elements equal to one into the sth group and for the zth c.c.d. is empty, we correspond only $u_{s,2}^z$ to $u_{s,w}^z, \forall w$ and $v_{s,2}^z$ to $v_{s,w}^z, \forall w$.

Corollary 5.4.3 (Loss cdf linked to the CHC_j) *The loss cdf linked to the CHC_j, with centroids $\mathbf{u} \in [0, 1]^{k_j}$, is the function $F_j^s : [0, n_j] \to [0, 1]$ such that*

$$F_j^s(i) = \sum_{l=0}^{i} \sum_{z=1}^{\hat{D}^c(l,k_j)} V_{CHC_j}(\mathbf{S}^z(\mathbf{p}_{l,k_j}(z))),$$

where $V_{CHC_j}(\mathbf{S}^z)$ is the volume of the CHC_j computed for the box $\mathbf{S}^z = [\mathbf{u}^z, \mathbf{v}^z] \in \mathbb{R}^a \times \mathbb{R}^a$, $\max(a) = 2k_j$, determined for the zth c.c.d. of l ones into k_j groups following the rule explained in Corollary 5.2.2, where $\mathbf{p}_{l,k_j}(z)$ is a $k_j \times \max(m_s^j, s = 1, \ldots, k_j)$ matrix of zeros and ones corresponding to the zth c.c.d. of l ones into k_j groups of dimensions m_s^j, $s = 1, \ldots, k_j$ and finally accounting for the permutation of the elements into the groups.

- The *between classes* aggregation step. As stressed before we consider a hierarchical copula structure for the dependence structure *between classes*, in order to set a hierarchical structure of risks. We consider n sub-portfolios, classified into the categories *large* or *small*. The *large* subclasses are basket of homogeneous risky loans that we consider affected by the same risk factor. We apply the limiting models to recover the loss pdf and cdf that we denote as $f_j^l(.)$ and $F_j^l(.)$, $\forall j = 1, \ldots, n_l$, respectively, where n_l denotes the cardinality of *large* classes. On the other hand, the dependence structure of the *small* subclasses is represented by a CHC, and then the elements are clusterized into homogeneous groups at a second level. We extract the loss pdf and cdf of the jth *small* class, that we denote $f_j^s(.)$ and $F_j^s(.)$, respectively, from the corresponding CHC_j, $\forall j = 1, \ldots, n_s$ where n_s stands for the cardinality of *small* classes.

  We are interested on determining the distribution of the whole loss defined as the sum of the classes' loss, $L = \sum_{j=1}^{n} L^j$,where $n = n_l + n_s$.
  Now the problem is to associate a counting variable to the copula function representing the dependence structure between the classes; this *first-level counting variable*, which represents the *whole loss cdf*, distributes the losses to the classes.

  Giving the vector of the loss cdf within classes

$$\mathbf{F}(\mathbf{q}) = \left(F_j^\varepsilon(q_j), j = 1, \ldots, n_l + n_s, (\varepsilon = s)\mathbf{1}_{n_j<20} \cup (\varepsilon = l)\mathbf{1}_{n_j \geq 20}\right),$$

  composed by limiting loss cdf for *large* classes and by a CHC-based loss cdf for *small* classes, we are interested on aggregating them with a dependence structure. We observe that the threshold to pass from a small to a large class is here assumed to be equal 20. By Sklar's theorem, being these functions uniformly distributed, we can represent their dependence structure as a copula $HYC(F_1^\varepsilon(q_1), \ldots, F_n^\varepsilon(q_n))$, where $n = n_l + n_s$. This is a hierarchical structure characterizing the highest or the first level of dependence.

We consider the default event and the corresponding counting variable on this event, that we call *first-level counting variable*. Here we assume to count some unit of percentage losses.

Definition 5.4.4 (First-level counting variable linked to a HYC function) The first-level counting variable linked to the $HYC(\mathbf{F}(\mathbf{p}_{l,n}(j)))$, where $\mathbf{p}_{l,n}(j)$ is the jth c.c.d.[29] of l units' percentage losses into n groups and $\mathbf{F}(\mathbf{p}_{l,n}(j))$ is the corresponding vector of margins computed through a limiting or CHC-based approach depending on the groups' cardinalities, is given by

$$wr(j) = \sharp(\mathbf{p}_{l,n}(j)), \forall j$$

which counts the elements of vector $\mathbf{p}_{l,n}(j)$ for the jth c.c.d.. Clearly we have $wr(j) = l, \forall j$.

Our aim is to recover the *whole loss cdf* or the cdf of the *first-level counting variable* linked to the n-dimensional copula function HYC. To do this we observe that a number l corresponds, through the whole loss cdf, to a number in $[0, 1]$; this pdf depends on all the c.c.ds of l ones into n groups (we enumerate this c.c.d. with the index j). The jth c.c.d., corresponding to the vector $\mathbf{p}_{l,n}(j)$, is related through a copula function, to a joint probability. This joint probability corresponds to the probability to have a loss less than or equal to l percentage units distributed into the groups as explained by the jth c.c.d.. The mean of the joint probabilities computed for all the c.c.ds for the same number of ones, i.e., for the same l ones into n groups, represents the (mean) cumulated probability to count l ones; hence this probability refers to the scenario of l default events whose probabilities are the marginal cdf computed within the classes and where we assume that the different ways to distribute the events are equally probable. Anyway we may consider an expectation coherent with to the assumption related to the affecting policy of the events.

Corollary 5.4.4 (Expected whole loss cdf linked to a HYC function) *The expected whole loss cdf linked to the n-dimensional first-level dependence HYC, characterized by n homogeneous groups of dimensions $m_s, s = 1, .., n$, respectively, such that $\sum_{s=1}^{n} m_s = N$ is the function $F^{wr} : [0, N] \to [0, 1]$ such that*

$$F^{wr}(i) = \mathbb{E}\left(C(\mathbf{F}(\mathbf{p}_{l,n}(j))), j = 1, \ldots, \hat{D}^c(i, n)\right) \tag{5.7}$$

where $\mathbb{E}$ is an expectation coherent with the affecting policy, $\mathbf{p}_{l,n}(j)$ is a n-vector counting the ones corresponding to the jth c.c.d. of l ones into n groups of dimensions $m_s, s = 1, \ldots, n$, $\mathbf{F}(\mathbf{p}_{l,n}(j))$ is the vector of margins corresponding

[29] Here the compatibility is with respect to the groups' cardinalities. These c.c.ds are ordered with respect to some specified criteria.

to the jth c.c.d. of l ones into n groups, and $\hat{D}^c(i, n)$ *counts the c.c.ds of i ones into n groups.*

We can observe that the suggested hybrid model is a not exchangeable model since the classes are represented by different default probabilities $p^j, j = 1, \ldots, n$. This property can be seen as a correction of the exchangeable one-factor models (see Gordy 10, 11).

Remark 5.4.1 We see that in order to define the right weights in (5.7), it's necessary to define how the defaults will distribute into the groups. If we assume that the groups are equally affected by the defaults' occurrence or that the probability that a default attaches a group i is the same of group $j, \forall i, j$, the expectation operator in (5.7) becomes the arithmetic mean.

5.4.3 *Check for the Groups' Cardinality: The HYC Model*

The limiting models are useful in order to diminish the complexity of the problem and in order to implement an hybrid model we need to identify which groups could be considered large. Here the implementation of the HYC is discussed with the detail of the MatLab™ code.

```
% We associate to every row of the c.d. matrix, numberOfDefaultsInGroup the
% right marginal selecting the right model depending on the group's cardinality

function [HCM] = HCVL2_hibrid(numberOfDefaultsInGroup,pDefVector,...
cardinalityOfGroups,alpha,alpha_0,rho,c)

for j=1: size(numberOfDefaultsInGroup,2)
% the cardinality distinguishes between small and large groups
% (given a threshold)
for i=1: size(numberOfDefaultsInGroup,1)
if cardinalityOfGroups(1,j)< threshold
for t=0:cardinalityOfGroups(1,j)
 % if small we apply the CHC with single groups (infinite granularity)
 % we compute the pdf
CVM(t+1,j)= VOL_CHC1(t,pDefVector(1,j),1,cardinalityOfGroups(1,j),alpha(1,j),...
alpha(1,j),1)
end
CVM1(:,j)=[CVM(:,j)];
% we compute the cdf, i.e. the margins
CVM2(:,j)=cumsum(CVM1(:,j));
HCM1(i,j)=CVM2(numberOfDefaultsInGroup(i,j)+1,j);
% if large, a limiting model is selected for the computation of margins
else
switch c
 % Limiting Clayton within the groups
    case{1}
HCM1(i,j)=clayton_limit_1(pDefVector(1,j),cardinalityOfGroups(1,j),...
numberOfDefaultsInGroup(i,j),alpha(1,j));
% Limiting Gaussian within the groups
   case{2}
HCM1(i,j)=gaussian_limit_1(pDefVector(1,j),cardinalityOfGroups(1,j),...
```

```
numberOfDefaultsInGroup(i,j),rho(1,j));
end
end
end
 end
 % the marginals are finally aggregated with the first level dependence
 % (here Clayton)
 for i=1: size(HCM1,1)
 for j=1:size(HCM1,2)
 HCM2(i,1)=copula_Clayton(HCM1(i,:),alpha_0);
 end
 end
 % we have a vector of probabilities related to the same number of ones but
 %distributed in different ways
HCM=[HCM2(:,1)];
%HCM is the mean default probability
HCM=sum(HCM)/size(HCM,1);
```

Example 5.4.4 (Concentration and Granularity in the HYC model) In order to assess the HYC model rating the improvements with respect to a PL model, we propose this exercise. In this example we consider a loan portfolio of dimension $N = 60$ clusterized[30] into $n = 3$ classes. These groups are assumed to be homogeneous with marginals equal to the centroids of the same group, which represents the vector of the default's probabilities:

$$Centroids = \left[0.035 \; 0.057 \; 0.074 \right]$$

The dependence parameters within the groups correspond to the following levels of Kendall τ: 0.6, 0.43, and 0.8, while the parameter between the groups corresponds to a Kendall $\tau = 0.0698$. We assume a Clayton dependence structure between the groups, meaning that the dependence parameter is $\alpha_{\text{between}} = 0.15$.

We compare the *whole loss cdf* computed in four scenarios characterized by different levels of concentration. The scenarios considered are the following:

1. *Equally distributed* case, corresponding to three groups, each one composed by 20 loans
2. *Not perfectly equally distributed* case, corresponding to a cardinalities' vector $n^2 = (10, 25, 25)$
3. *Concentration on the riskiest class*, corresponding to a cardinalities' vector $n^3 = (5, 5, 50)$
4. *Concentration on the safest* class, corresponding to a cardinalities' vector $n^4 = (50, 5, 5)$.

[30]We assume to group with a clustering method, based on the Euclidean metric with a little tolerance, as $\delta = 0.09$.

We compare the mean *whole loss cdf* computed following the HYC approach for Gaussian and Clayton limiting (Gaussian/Clayton HYC for short) with the PL model corresponding to the case of a large number of *large* classes.[31]

In Fig. 5.24a, b, we compare the *whole loss cdf* in the HYC Clayton and Gaussian model, respectively, and the corresponding PL model having the same Kendall τ into the highest level of dependence, and we consider different scenarios. We observe that the Clayton PL model compared to the HYC model approximates better the mean *whole loss cdf* in the *equally distributed* case and the goodness of approximation decreases with the increasing of the concentration toward the riskiest class. This error is composed by a *clusterization error*, including the *exchangeable error*, which corresponds to the *sector aggregation risk*, and the *equally affecting policy error*. A second component of the error is given by a *partial error of granularity*.[32] This last kind of error is avoided into the HYC model since here we consider a not perfectly fine-grained assumption, or we recover the fine grain into the *small* classes. The impact of the errors is greater for lower level of percentage losses and generally greater for the Gaussian PL model than for the Clayton one. On the other hand, the *clustering error* depends to the fact that the PL models do not specialize the kind of distribution considered, since the loans are assumed exchangeable and equally affected, i.e., are not clusterized into groups.

Figure 5.24c, d explain the same comparison as before but for a different level of Kendall τ. We observe that the error of PL cdf is lower for higher level of Kendall τ and that the PL Clayton approximates better the *equally distributed* case, for a lower level of dependence, and the concentrated case, particularly toward the riskiest class, for an higher level of dependence.

Figure 5.25a, b compare the Clayton HYC and the Gaussian HYC mean *whole loss cdf* with the corresponding cdf computed through the PL model for a particular concentration scenario and for several dependence levels. We observe that increasing the correlation between classes, the HYC loss cdf goes toward the PL loss cdf where we have a concentration on the same class. The *clusterization error* of PL models becomes lower for higher dependence between the clusters.

Figure 5.25c, d compare the Clayton/Gaussian HYC mean *whole loss cdf* with the Clayton/Gaussian HLC mean *whole loss cdf* and the corresponding PL loss cdf for different concentration scenarios. We observe that into the analyzed scenarios the HLC cdf suffers of a big *partial error of granularity* that is comparable to the approximation of the PL cdf. This implies that in this case the *clusterization error* is quite non-influential. The scenarios analyzed give the same information, and the evidences are invariant for the level of dependence considered.

---

[31] We observe in the first scenario the mean *whole loss cdf* computed with the HYC approach coincides with the HLC mean *whole loss cdf* since in this case all classes are *large*.

[32] We talk about a *partial error of granularity* since we compare the PL with the HYC whole loss cdf that corrects the granularity only for *small* classes. The properly *error of granularity* can be recovered by a comparison between the PL and the CHC whole loss cdf.

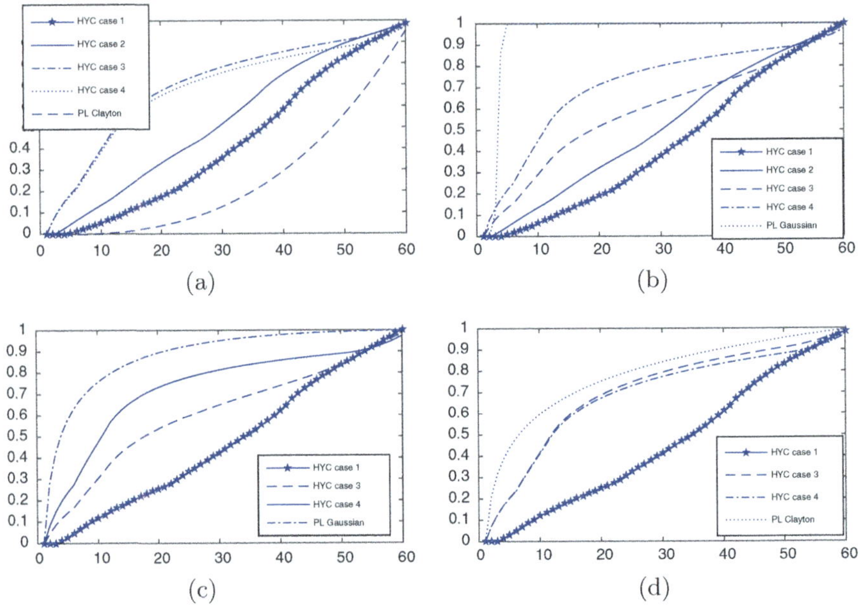

Fig. 5.24 Mean *whole loss cdf: HYC versus PL models.* (**a**) Clayton HYC mean *whole loss cdf* for several concentration scenarios vs PL Clayton loss cdf. Kendall $\tau = 0.0698$ for PL and the highest level of Clayton HYC model. (**b**) Gaussian HYC mean *whole loss cdf* for several concentration scenarios vs PL Gaussian loss cdf. Kendall $\tau = 0.0698$ for PL and the highest level of Gaussian HYC model. (**c**) Gaussian HYC mean *whole loss cdf* for several concentration scenarios vs PL Clayton loss cdf. Kendall $\tau = 0.43$ for PL and the highest level of HYC model. (**d**) Clayton HYC mean *whole loss cdf* for several concentration scenarios vs PL Clayton loss cdf. Kendall $\tau = 0.43$ for PL and the highest level of HYC model

In Fig. 5.26a we plot the mean *whole loss cdf* for Clayton HYC, Clayton HLC, and PL Clayton models for the *not perfectly equally distributed* case and for several levels of Kendall τ. We observe that the PL cdf explains a *clusterization error* that is dominant with respect to the *partial error of granularity*. Moreover the *partial error of granularity* decreases if the dependence level between groups increases.

Figure 5.26b, d compare Clayton and Gaussian HYC and Clayton and Gaussian HLC mean *whole loss cdf*, respectively, in several concentration cases. The evidence shows that the *partial error of granularity* increases with the concentration, independently of the direction of this concentration,[33] since the number of *small* classes increases.

In Fig. 5.26d we compare the Gaussian HYC and the Gaussian HLC mean *whole loss cdf* with the clusterized homogeneous Clayton copula-based (CHC for short) loss cdf (see Bernardi and Romagnoli 4) that we consider as our benchmark, in the

[33] This means that the evidence explained here is true if we consider a concentration toward the riskiest class but also toward the safest one.

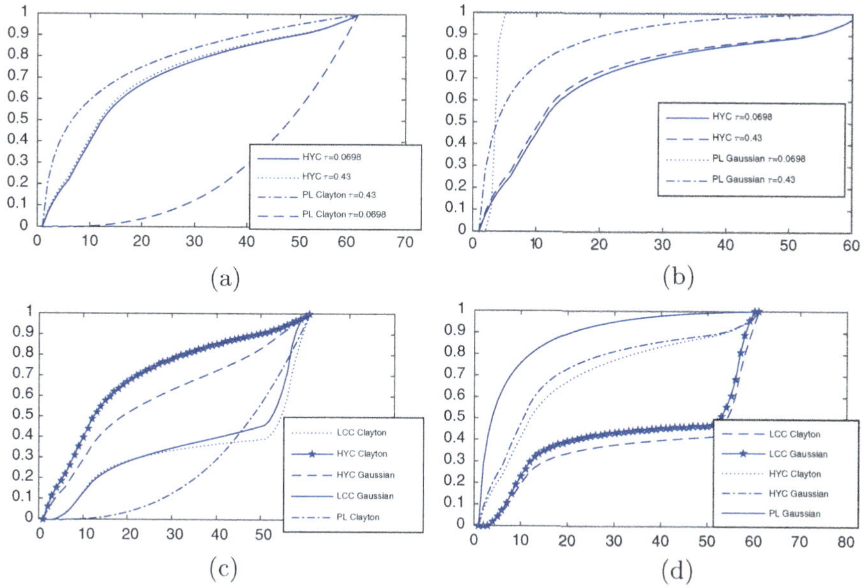

Fig. 5.25 Mean *whole loss cdf: HYC versus PL models.* (**a**) Clayton HYC mean *whole loss cdf* vs PL Clayton loss cdf in the concentration on the riskiest class case. Several levels of Kendall for PL and the highest level of HYC model. (**b**) Gaussian HYC mean *whole loss cdf* vs PL Gaussian loss cdf in the concentration on the riskiest class case. Several levels of Kendall for PL and the highest level of HYC model. (**c**) Clayton/Gaussian HYC vs Clayton/Gaussian HLC mean *whole loss cdf* and PL Clayton loss cdf in the concentration on the riskiest class case. Kendall $\tau = 0.0698$ for PL and the highest level of HYC model. (**d**) Clayton/Gaussian HYC vs Clayton/Gaussian HLC mean *whole loss cdf* and PL Gaussian loss cdf in the concentration on the safest class case. Kendall $\tau = 0.0698$ for PL and the highest level of HYC model

not perfectly equally distributed scenario. The difference between the HLC mean *whole loss cdf* and the CHC loss cdf represents the properly *error of granularity* since into the HLC model all classes are considered *large*, i.e., with an infinity degree of granularity. On the other hand, the difference between the HYC mean *whole loss cdf* and the CHC loss cdf explains an error of granularity attaching only the classes classified as *large*; we call this error as *semi-granularity error*. The scenario considered here complains only one small class, implying that the improvement of HYC with respect to the HLC, i.e., the *partial error of granularity*, is small. We observe also that the CHC loss cdf is better approximated by the HYC cdf for the concentrated scenarios (toward the riskiest or the safest class) since in this case the HYC model corrects the granularity of two classes. Anyway we may conclude that if our benchmark is the CHC model in the *not perfectly equally distributed* scenario, the granularity aspect seems to be prevailing on the concentration assumptions.

Finally we discuss the scalability of the suggested algorithm to compute the mean *whole loss cdf* with the HYC model, i.e., the maximum dimension to

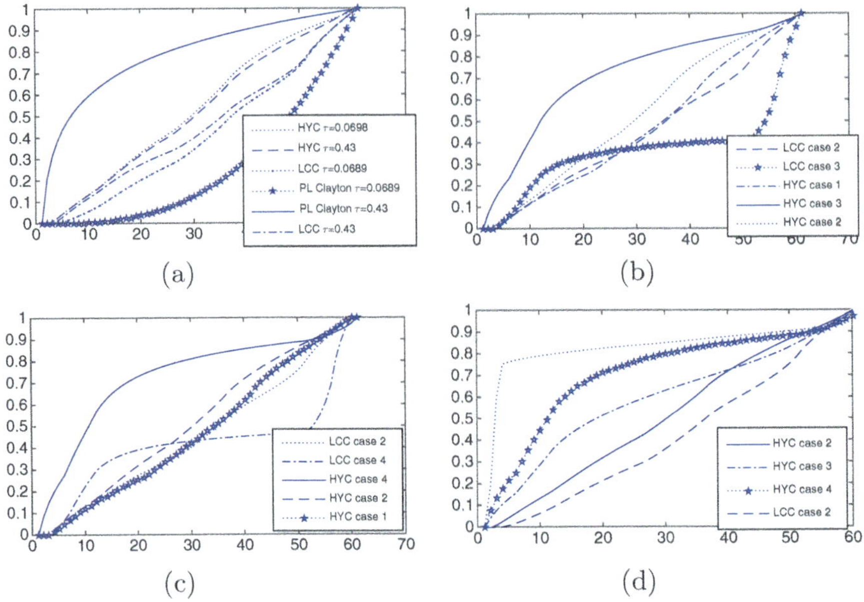

Fig. 5.26 Mean *whole loss cdf: HYC versus PL, HLC, and CHC models.* (**a**) Clayton HYC vs Clayton HLC mean *whole loss cdf* and PL Clayton loss cdf in the *not perfectly equally distributed* case. Several levels of Kendall τ for PL and the highest level of HYC model. (**b**) Clayton HYC vs Clayton HLC mean *whole loss cdf* in several concentration scenarios. Kendall $\tau = 0.43$ for the highest level of HYC and HLC models. (**c**) Gaussian HYC vs Gaussian HLC mean *whole loss cdf* in several concentration scenarios. Kendall $\tau = 0.43$ for the highest level of HYC and HLC models. (**d**) Clayton HYC and Clayton HLC mean *whole loss cdf* vs CHC in the *not perfectly equally distributed* case. Kendall $\tau = 0.048$ for the highest level of HYC, HLC and CHC models

which our techniques can be extended. We analyze the scalability of the HYC algorithm with respect to the number of groups and the level of concentration. The MatLab™ algorithm was launched on a MacBook Pro, 2,8 GHz Intel Core i7 quad-core, 16 GB 2133 MHz LPDDR3. Table 5.3 reports the computing time for the first 10 percent of the *whole loss cdf* corresponding to the sample of dimension $n = 60$ for different numbers of groups and concentration assumptions and the CPU time-saver of the HLC and the PL approaches with respect to the HYC method. We consider three different concentration scenarios for the case of three clusters: (i) the equally distributed scenario for the cardinalities $(20, 20, 20)$ performs three *large* classes; (ii) the not perfectly equally distributed case for cardinalities $(10, 25, 25)$ or with two *large* and a *small* class; and (iii) a concentration case for cardinalities $(5, 5, 50)$ or for two *small* and a *large* class. We observe that into the equally distributed scenario, the HYC model coincides exactly with the HLC one since all classes are large implying that the *partial error of granularity* overlaps the *semi-granularity error* attaching the HYC procedure.

Table 5.3 CPU computation time in seconds and CPU time-saver with respect to the Clayton HYC algorithm of the left queue (till 10 percent) of the mean *whole loss distribution* with HLC and PL approaches. The dimension of the set is $n = 60$

N. groups/conc.	Cpu time	Ts Gaussian/Clayton HLC	Ts PL Gaussian/Clayton
2/eq.dist.	0.23	0.017/0.02	0.16/0.21
3/eq.dist.	1.58	0.11/1.01	1.52/1.2
3/np eq.dist.	56.81	61.15/62.21	62.58/63.71
3/conc.	31.8	30.16/31.19	33.65/31.52

Table 5.4 Percentage absolute errors of the loss distributions' deciles computed with different models with respect to the CHC-based whole loss distribution in the equally distributed scenario. Dimension of the set $n = 60$ and Kendall $\tau = 0.43$

Model	AE 3/eq.dist.	AE 3/np eq.dist.	AE 3/conc.
Clayton HYC	24.01%	25.31%	4.51%
Gaussian HYC	23.44%	22.72%	11.25%
Clayton HLC	22.01%	25.56%	22.45%
Gaussian HLC	17.64%	21.95%	22.35%
PL Clayton	25.7%	24.7%	26.2%
PL Clayton	31.21%	31.31%	32.11%

We can observe that the number of groups is a critical variable; the CPU time increases exponentially with the number of group. Indeed we observe that HLC and PL approaches are faster than HYC algorithm; the CPU time-saver corresponds to the price to correct a *partial error of granularity* as it is shown in Table 5.3. Here we compare the relative errors of the deciles of the *whole loss cdf* computed with different methods with respect to our benchmark that is the CHC cdf representing the best approximations for high-dimensional problem (see Bernardi and Romagnoli 2) for a Kendall $\tau = 0.43$. We notice that when the number of *small* groups increases, the performance of the HYC method is clearly better than other models (Table 5.4).

References

1. Barlow, R. E., & Proschan, F. (1975). *Statistical Theory of Reliability and Life Testing: Probability Models*. Holt: Rinehart and Winston.
2. Bernardi, E., & Romagnoli, S. (2011). Computing the Volume of an high-dimensional semi-unsupervised Hierarchical copula. *International Journal of Computer Mathematics, 88*(12), 2591–2607.
3. Bernardi, E., & Romagnoli, S. (2012). Limiting Loss distribution on a Hierarchical copula-based model. *International Review of Applied Financial Issues and Economics, 4*(2), 126–145.
4. Bernardi, E., & Romagnoli, S. (2013). A Clusterized copula-based Probability Distribution of a counting variable for high-dimensional problems. *The Journal of Credit Risk, 9*(2), 3–26.
5. Bernardi, E., & Romagnoli, S. (2015). A copula-based hierarchical hybrid loss distribution. *Statistics and Risk Modeling, 32*, 73–87.

6. Bernardi, E., & Romagnoli, S. (2016). Distorted copula-based probability distribution of a counting hierarchical variable: A credit risk application. *International Journal of Information Technology & Decision Making, 15*, 285–310.
7. Cifuentes, A., & O'Connor, G. (1996). The binomial expansion method applied to CBO/CLO analysis. *Moody's Special Report*, December.
8. Cifuentes, A., Efrat, I., Glouck, J., & Murphy, E. (1999). Buying and selling credit risk: a perspective on credit-linked obligations. In J. Gregory (Ed.). *Credit Derivatives*. London: Risk Book, pp. 112–123.
9. Emmer, S., & Tasche, D. (2006). Calculating credit risk capital charges with the one-factor model, *Journal of Risk, 7*, 85–101.
10. Gordy, M. (2003). A risk-factor model foundation for ratings-based bank capital rules. *Journal of Financial Intermediation, 12*(3), 199–232.
11. Gordy, M. (2004). Granularity adjustment in portfolio credit risk measurement. In G. Szego (Ed.), *Risk Measures for the 21st Century*. Chichester: Wiley.
12. Hofert, M., Mächler, M. (2011). Nested Archimedean Copulas meet R: The nacopula package. *Journal of Statistical Software*, in press.
13. Joe, H. (1997). *Multivariate Models and Dependence Concepts*. London: Chapman & Hall.
14. Lindskog, F., & McNeil, A. J. (2003). Common Poisson shock models: applications to insurance and credit risk modelling. *ASTIN Bulletin, 33*(2), 209–238.
15. Marshall, A. W., & Olkin, I. (1967a). A generalized bivariate exponential distribution. *Journal of Applied Probability, 4*, 291–302.
16. Marshall, A. W., & Olkin, I. (1967b). A multivariate exponential distribution. *Journal of American Statistics Association, 62*, 30–44.
17. Marshall, A. W., & Olkin, I. (1988). Families of multivariate distributions. *Journal of the American Statistical Association, 83*, 834–841.
18. Mehta, M. L. (1991). *Random Matrices*, 3rd edn. New York: Academic.
19. Schönbucher, P. (2003). *Credit Derivatives Pricing Models: Model, Pricing and Implementation*. Chichester: Wiley.
20. Schönbucher, P. J. (2004). *Taken to the Limit: Simple and Not-so-simple Loan Loss Distributions*, Wilmott magazine.
21. Vasicek, O. (1987). Probability of loss on loan portfolio. *Working paper*, KMV Corporation.
22. Vasicek, O. (1991). Limiting loan loss distributions. *Working paper*, KMV Corporation.
23. Vasicek, O. (1997). The loan loss distribution. *Working paper*, KMV Corporation.
24. Vasicek, O. (2002). Loan portfolio value. *Risk, 15*(2), 160–162.
25. Wang, L., Bo, L., & Jiao, L. (2006). A modified K-means clustering with a density-sensitive distance metric. Rough sets and knowledge technology. *Lecture Notes in Computer Science, 4062*, 544–551.

Chapter 6
Real Data Empirical Applications

Abstract The goal of this chapter is to display a number of empirical applications of some of the new aggregation techniques discussed in the book. We give details of several examples, from risk management to portfolio selection, where market data are used to properly calibrate the dependence structure and to identify the more suitable approach to the problem at hand.

6.1 HYC-Based Model for a Worldwide Sovereign Debt Large Portfolio

In order to emphasize how relevant the dependence structure linked to the worldwide economy is for the valuation of the sovereign risk that must be considered so as to avoid over/under valuation of such same risk, we structure a large portfolio exposed to the worldwide sovereign debts. This analysis is inspired by Bernardi et al. [3].

All data are drawn from Thomson Reuters Datastream. We choose 59 countries representing the worldwide economy, and we construct a data set composed by the CDS prices (for eight relevant maturities[1]) at the 9th of February 2013.

Starting from CDSs' prices, we get implicit default probabilities and compute them by a recursive bootstrapping method, i.e., through the following formula

$$\mathbb{Q}(n) = \mathbb{Q}(n-1)\left[1 - \frac{CDS(n)}{LGD}\right] - \frac{CDS(n) - CDS(n-1)}{B(0,n)LGD}\sum_{i=1}^{n-1} B(0,i)\mathbb{Q}(i-1)$$

where $\mathbb{Q}(n)$ is the survival probability at time n, $B(0, n)$ stands for the discounted function from n to 0, $CDS(n)$ is CDS price with maturity at time n, and LGD is the *loss given default* that we fixed at 0,5 value. Therefore the bootstrapped default

[1] We consider eleven maturities totally: eight of these come from real market data, while the other corresponding to a maturity of 6, 7, and 8 years are interpolated prices.

E. Bernardi, S. Romagnoli, *Counting Statistics for Dependent Random Events*,
https://doi.org/10.1007/978-3-030-64250-1_6

probabilities are recovered and will be our first input to be utilized in clustering the countries.[2]

Our first goal is to compute the loss distribution of this portfolio. To this aim we reduce the complexity of the problem, applying a clustering method to the data set. We employ the self-organizing map (SOM in the following) to clusterize our sample, considering as features the bootstrapped default probabilities and an index of the country's risk. The index used in the following analysis is the SACE's global market risk that is a country risk index understood as a mix of sovereign, political, economic, money transferring, and location risks. It is made up by two scores: a credit risk score and a market risk score. The credit risk score is calculated starting from financial situations (foreign debt/money income, foreign debt/GDP, etc.) and then taking into account economic situations (growth potential, policy performance, shock resistance), payment expertise and finally adjusted with political situation. This evaluation leads to a credit score that assesses the country credit risk with a value in a range from 0 to 0.7 where 0 represents the minimum risk. On the other hand, the market risk score is a combination of political risk (internal policies and international relationships), economic risk (economic activity, public balance and inflation, GDP, exchange rate, domestic market), banking system risk (structure and performance), and operating risk (legal system, security, etc.); each of the previously listed risks gives rise to a risk/outlook combination that is synthesized in a score defined in a range from 0.1 to 0.9. A weighted average of credit risk and market risk scores translates into the GMR index: this is a value defined in a range from 0.1 to 1 that goes from L1 (lower risk with score 0.1) to H3 (maximum risk with score 1). Countries' risk indices are inserted in Table 6.1.

SOM is a competitive layer for classification problem that is optimized with four groups for all the maturities, represented by one large sub-portfolio and three small ones, for a maximum *mean confusion* of 2.75×10^{-1}.

A very important role was played by correlation parameters within and between groups that have been evaluated based on the time series of CDSs' prices from the 14 December 2008 to the 9 February 2013, respectively, for all the maturities. To compute them we applied the Genest and Rivest's [1993] procedure that is detailed in Frees and Valdez [7]. Given a set of observations $(X, Y)_i, i = 1, \ldots, n$, this procedure allows us to identify the Archimedean copula that best represents the distribution function of (X, Y), in terms of the corresponding mean square error minimization.

To estimate the dependence parameters within the groups, we did a second level of clustering of each group in order to divide it into sub-groups. For each sub-group we then compute an average time series of CDSs' prices, and, using Genest and Rivest's procedure, we calculate the correlation parameters within the groups. The same procedure, which works with the average time series of CDSs'

---

[2]Given the anomalies on the CDSs' prices of the Greece during the analyzed period, we assume a flat structure of CDS price equal to the asset swap spread of the Hellenic Republic step-up coupon bond (isin: GR0128011682) at the 24th of January 2013 and with 12 years maturity.

Table 6.1 GMR indexes

Country	GMR	Country	GMR	Country	GMR
Argentine	1	Iceland	0.7	Pilipines	0.7
Australia	0.1	Indonesia	0.6	Poland	0.4
Bahrain	0.4	Iraq	0.9	Portugal	0.7
Belgium	0.3	Ireland	0.6	Qatar	0.3
Bolivia	0.9	Israel	0.4	Romania	0.6
Brazil	0.4	Italy	0.4	Russia	0.6
Bulgaria	0.7	Jamaica	1	Serbia	0.6
Chile	0.3	Japan	0.2	Singapore	0.4
Colombia	0.5	Kazakhstan	0.8	Slovak Republic	0.5
Croatia	0.7	Korea	0.3	Slovenia	0.6
Cyprus	0.8	Latvia	0.6	South Africa	0.4
Czech Republic	0.5	Lebanese Republic	0.8	Spain	0.5
Denmark	0.1	Lithuania	0.6	Sweden	0.2
Dominican Republic	0.9	Malaysia	0.5	Thailand	0.5
Estonia	0.5	Malta	0.4	Tunisie	0.8
Finland	0.1	Mexico	0.5	Turkey	0.6
France	0.3	Morocco	0.6	Uk and Northern Ireland	0.3
Germany	0.2	Norway	0.1	Ukraine	0.9
Greece	0.9	Panama	0.5	Uruguay	0.7
Guatemala	0.9	Peru	0.5		

prices of each group, enables us to recover the dependence parameters between the groups. Clayton copula is the Archimedean copula that performs at best under the Genest and Rivest [9] procedure in terms of minimum square error (MSE for short). Cardinalities of groups generated by SOM clustering, default probabilities and GMR indices centroids such as Kendall's tau within and between and the corresponding Clayton dependence parameters with their estimated mean square errors, are evaluated. We point out that the correlation within groups is always positive except for the couple Argentine and Cyprus, representing the first group at maturity 3Y; they are aggregated in the same group because of the similarities in their default probabilities and GMR indexes, but CDSs' prices time series, from which we estimate the dependence parameters, show an opposite behavior in time.

We compute the whole portfolio's loss function through the HYC model which selects the Clayton limiting model (PL Clayton) or the Gaussian one (PL Gaussian) for large groups; otherwise it proceeds by using the CHC method. Figure 6.1 shows the loss surfaces specialized for short term (from 6M to 2Y) and long term (from 6Y to 10Y) both in Clayton and Gaussian case. We notice how long-term loss surfaces are the most regular ones, while other surfaces, especially short-term surfaces, present anomalies perhaps due to contingent facts.

In Fig. 6.2 we compare the loss distribution for several maturities (2Y-5Y-10Y) both in the Clayton and Gaussian cases: in subplots Fig. 6.2a, b, it would seem that a

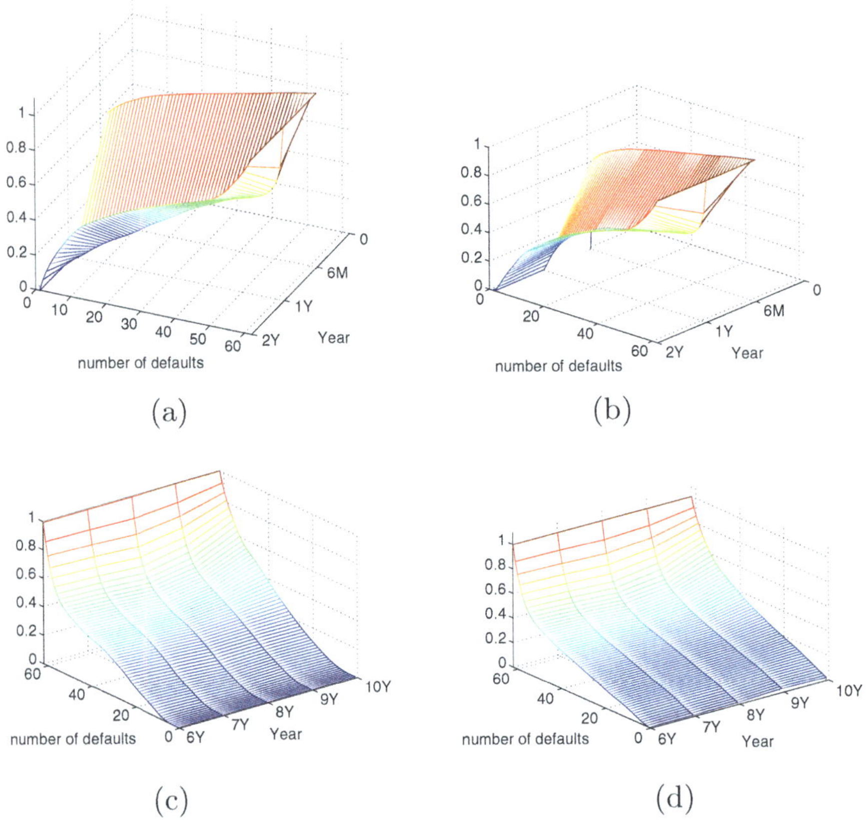

Fig. 6.1 Loss surfaces: Clayton and Gaussian dependence. (**a**) Short-term loss surface: Clayton dependence (**b**) Short-term loss surface: Gaussian dependence. (**c**) Long-term loss surface: Clayton dependence. (**d**) Long-term loss surface: Gaussian dependence

less conservative policy could correspond to a longer maturity, because shorter-term distributions dominate longer ones.

Furthermore we compute the loss function through the HLC approach where all groups are considered large and then request a limiting approach. We report several evidences coming from a comparison of the HYC loss function and the corresponding HLC and the limiting PL distribution in Clayton case for short and long maturities. The distance between HYC and HLC curves represents the partial error of granularity. Here this error involves three to four groups because only one group is large, and then it is not corrected for granularity. The results show how the partial error of granularity is decreasing with respect to maturity and very large for short-term curves in the Clayton case. On the other hand, it appears moderated and quite stable with respect to maturity in the Gaussian case. Moreover HLC Gaussian model seems to overrate HYC curve at high default percentage (more than 40 defaults), maybe because assets with high default probabilities have an important

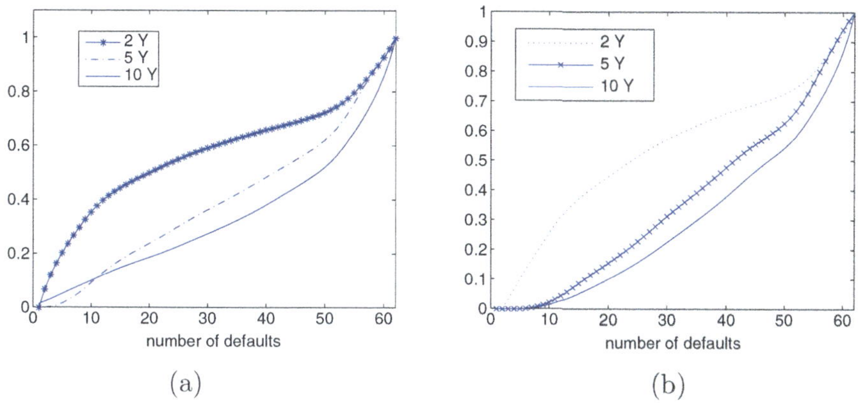

Fig. 6.2 Loss distribution for several maturities: Clayton and Gaussian dependence. (**a**) Loss distribution for several maturities: Clayton dependence (**b**) Loss distribution for several maturities: Gaussian dependence

weight on the infinite granularity portfolio. On the other hand, Clayton HLC model seems to show a better estimate of the loss HYC function for longer maturity, where a lower tail dependence model, which frequently characterizes financial data, seems to fit better than a symmetric one, strictly connected with normal distribution. Obviously HYC method should give the best approximation of the portfolio's loss function with respect to HLC, due to the partial correction of granularity in small groups.

It is also interesting to compare HLC loss functions and the corresponding PL curves, computed considering a homogeneous structure of the default probabilities, taken equal to the average portfolio's default probability, and a dependence structure equal to the first level Kendall's τ of the clustering method. It is evident that the pure limiting cdfs are corrected by the HLC distributions which recover the sector concentration considering several classes of risk. PL Clayton loss function seems to suggest a less cautious behavior that is directly proportional to the correlation level as pointed out in [2]. In the Clayton case, concentration error appears very small for short and long terms, while it is bigger for long-term curve where PL distribution seems to better approximate HYC for a compensative effect of granularity and concentration errors. On the other hand, concentration errors are dominant and very high in Gaussian case where PL curve shows an extreme prudential attitude; the correction proposed by HLC is a good approximation of HYC curve since the impact of granularity error is residual with respect to the one of concentration. Moreover the granularity error seems to be larger for long maturity.

6.2 Risk Evaluation Based on HYC Model: A Credit-Exposed European Investment Portfolio Analysis

In order to evaluate the credit risk under which financial institutions are submitted, we define a portfolio composed by 55 CDS from 7 different European countries inspired to Bernardi et al. [5]. In Table 6.2 the set of chosen CDS are listed; it is clear that the portfolio is medium diversified since there are 4 different industries involved.

6.2.1 *Copula-Based Loss Distribution*

We consider the previously defined portfolio of credit risk exposed assets that we assume clusterized into a finite number n of sub-portfolios or classes affected by different risk factors. The basic idea is to introduce a hierarchical risks structure allowing to take into account the non-exchangeability of the variables and the grains of such classes. It's important to highlight that we take into account the sector concentration, but we consider the right granularity only for groups that

Table 6.2 Credit-exposed European investment portfolio with evidence of the geographical exposure

France	Germany	Italy	Netherlands
Air France	Adidas	Banca Monte dei Paschi	Assa Abloy
Alcatel – Lucent	Allianz	Banca Popolare di Milano	Atlas Copco
Axa	Bayer	Edison	Aegon
BNP paribas	BMW	Enel	Electrolux
Carrefour	Commerzbank	Fiat	Heineken
Danone	Continental	Finmeccanica	PostNL
EDF	Daimler	Intesa San Paolo	
L'Oreal	Deutsche bank	Mediobanca	
Michelin	Deutsche telekom	Pirelli	
Peugeot	Linde	Seat Pagine Gialle	
Spain	**Switzerland**	**United Kingdom**	
Banco de Sabadell	Adecco	3i Group	
Banco popular espanol	Clariant	Abbey	
Endesa	Credit Suisse	Anglo American plc	
Gas natural	Nestle	Barclays	
Iberdrola	Roche holding	British American tobacco	
Melia hotel		Cable & wireless	
Repsol		Compass group	
Telefonica		Daily mail	

are deemed *small*,[3] while for greater cardinalities we use a limiting approach to represent the dependence structure within the groups. The implementation of the HYC methodology is concerned basically with the reduction of the complexity of the problem with respect to the dimension of the data set. This may be reached by a clustering and a homogenization procedure.

In the empirical application, we opted both at the first and at the second level for a semiunsupervised SOM[4] based on the Euclidean metric and where the number of clusters is optimal in term of a minimization of the clustering error defined as a sum of *confusions* (see Wang et al. 16). Moreover we recall that in Bernardi and Romagnoli [1], the homogenous assumption is compensated with a change of groups' cardinality of the clusterized copula function that is called *equivalent* in this sense. Therefore here we don't change the group's cardinality, but we control the homogenization error through an optimal choice of the number of clusters in terms of a given *total confusion* at both the first and the second grouping level. More precisely our purpose was to divide the sample in groups as much homogeneous as possible, fixing a discriminant variable. After careful consideration we recognized the daily average variation (DAV in short) of the equity as the most efficient variable in order to find a first clusterization into four principal groups. Subsequently we proceeded with a second step grouping, by dividing the previous four groups in two sub-groups each, based on the default probabilities (PD for short). The passage described above has been accomplished through a competitive layer for classification problem called SOM, which is composed by a topology function calculating the neurons positions in an hexagonal pattern and a distance function (that in our application is the Euclidean distance) used to find distance between layer's neurons given their positions. SOM is a single layer characterized by a weight function and an input function corresponding to the vectors of the bootstrapped default probabilities and the DAV variables. The layer receives a weight (that is randomly initialized) from the input and the processes of adaptation and training update the weights through the learning function. Training occurs until a maximum number of iterations (fixed on 1000 in this application) or until the performance goal is met.

The information that we used to cluster are recovered from market data.[5] Particularly we recovered the PD through the Bootstrapping procedure; therefore starting from CDSs' prices and by equating the protection and the premium legs with a recursive forward procedure, we got implicit default probabilities assuming a loss given default fixed at 0.6 value (being the Recovery Rate at 0.4 level).

[3] We need to fix a bound to allow for the classification of *small* group. Coherently with Schönbucher [15], we set a bound of 20.

[4] The acronym stands for self-organizing map which refers to a particular type of artificial neural network proposed by Kohonen [11].

[5] All data are drawn from Thomson Reuters Datastream.

6.2.2 Calibration of the Dependencies

Once all groups and sub-groups are defined, it is possible to proceed in the computation of copula parameters of our model. A very important role is in fact played by dependence parameters *within* and *between* groups that must be estimated based on a data set composed by 55 equities time series which goes from the 27 February 2007 to the 15 February 2014 (almost 7 years), respectively, for all the 10 maturities. For this purpose, we choose to work with Archimedean copula functions allowing us to reduce the definition of a multivariate copula to a single univariate function. In this class we select three families of Archimedean copulas, i.e., Clayton, Gumbel, and Frank, identifying a subclass that comprehensively covers all possible tail dependence assumption.
To calibrate the dependence structure, we applied the Genest and Rivest's [1993] procedure that is detailed in Frees and Valdez [7]. Given a set of observations $(X, Y)_i$, with i that goes from 1 to n, this procedure permits to identify different outcomes that are reported in Tables 6.3 and 6.4.
Finally a goodness of fit consideration allows us to select the copula function that is the "closest" to the empirical copula. In the following we took into consideration the Clayton copula parameter which seems to be the Archimedean copula performing the best according to Genest and Rivest procedure (in terms of the corresponding mean square error minimization).
Having computed the input variables for the HYC model, we then applied the algorithm yielding us the outcome represented by the loss cdf for every maturity.
From the graph we observe that there is a change in the trend after the first 5 years maturities. Indeed after the 5th year, the default probabilities go slower toward the end of the distribution, probably because the weight of probabilities is more equally distributed in the maturities which go from 6 to 10. The first five distributions (years from 1 to 5) reach 100% faster, and this means the probabilities of default are higher at the beginning: it is then more likely that we incur in a few number of defaults (the more time we have until the maturity and the more are the defaults that may occur in the remaining period). For instance, it is really hard to have 50 defaults in a portfolio of 55 CDS: it is clearly more likely to have them in a 10 years span.

Table 6.3 First-level copulas' parameters for *between* dependency. Since the first-level clusterization is the same, the correlation parameters reported here are the same for all the maturities

	Gumbel	Clayton	Frank
A – B	2.5487	3.0799	8.1782
B – C	4.8344	7.7088	17.1825
A – C	3.4128	4.5856	11.9904
A – D	1.2365	1.3830	4.0763
B – C	1.1094	0.4288	1.6161
B – D	1.6839	0.8529	2.2840

Table 6.4 Second-level copulas' parameters for the *within* dependencies: several maturities

			1Y	2Y	3Y	4Y	5Y	6Y	7Y	8Y	9Y	10Y
within - group	A1 - A2	Gumbel	1.51659	1.96766	1.99560	1.993460	1.92860	2.02590	2.09890	2.003790	1.06343	1.08743
		Clayton	1.02547	1.98433	1.93220	1.99450	1.69520	2.05480	2.04380	2.08680	0.12936	0.12526
		Frank	3.33476	5.72536	5.744534	5.72534	5.352534	5.764944	5.74935	5.74943	0.51235	0.51835
	B1 - B2	Gumbel	1.34347	1.24658	1.24458	1.27458	1.224658	1.27658	1.24968	1.243358	1.22558	1.21258
		Clayton	1.043394	0.49956	0.493716	0.47716	0.49256	0.95316	0.43316	0.42716	0.43716	0.44416
		Frank	3.43830	1.833680	1.83550	1.83480	1.82680	1.763780	1.83380	1.89380	1.82280	1.73780
	C1 - C2	NumOss										
		Gumbel	5.03630	3.92241	2.63649	2.67749	1.42702	1.42602	1.54102	1.44102	1.47102	1.41102
		Clayton	8.07330	5.82782	3.26397	3.25297	0.81104	0.84464	0.84204	0.84204	0.82204	0.83204
		Frank	18.34857	13.77163	8.45357	8.48847	2.86564	2.87544	2.86764	2.84364	2.78764	2.87564
	D1 - D2	Gumbel	1.50434	2.07266	2.04566	2.05566	2.80352	2.45095	2.40795	2.46595	2.47695	2.44595
		Clayton	1.019827	2.13433	2.145533	2.148633	3.60344	2.898189	2.86489	2.89889	2.88439	2.88689
		Frank	3.346483	6.04461	6.08661	6.06451	9.19806	7.67645	7.676534	7.67687	7.67579	7.67869

6.2.3 HYC Model: Portfolio Application

The application of the HYC model can be seen as a way of improving the computation of the risk of a portfolio considering the dependence structure implied by market data and evaluating every possible scenario based on the bootstrapped probabilities of default. Now the goal is to compute the VaR at a certain confidence level for which the calculation of the *expected* and *unexpected loss* is needed. If we have more than one scenario each one corresponding to a number of default i, for $i = 0, \ldots, n$, we know that they can be summarized together in a mean cdf that is the *whole loss* cdf introduced before.

As the Basel Committee writes on international regulations, *expected loss* is a function of the default probability, loss given default (LGD for short), and the exposure at default (EAD for short), i.e.,

$$EL_i = \Delta F^{wr}(i) \times LGD_i \times EAD_i,$$

where $\Delta F^{wr}(i)$ stands for the mean probability of the ith default scenario, for $i = 0, \ldots, n$.

Then it is possible to construct a cdf for EL, i.e.,

$$cEL_n = \sum_{i=0}^{n} EL_i.$$

On the other hand, the average computed over this distribution is the *expected loss* (EL_{mean} in the following) which indicates the amount that should be covered by capital requirement. If we want to consider $x\%$ confidence level, we know that, after this level, losses are uncovered. The area between this point and the mean is considered as the *unexpected loss*. The $x\%$ percentile of cEL corresponds to the total VaR of the portfolio at $x\%$ confidence level that we note as $TVaR_{x\%}$ in the following. To compute the *unexpected loss*, the relation is the following:

$$UL_{x\%} = TVaR_{x\%} - EL_{mean}.$$

6.2.4 HYC-VaR versus CM-VaR: an Empirical In-Sample Experiment

This experiment aims to empirically state a comparison between the credit metric approach (CM for short) and HYC models in the context of a risk management problem. We consider the short-term case, depending on the maturities of products composing the portfolio.

Table 6.5 Spot rates for several ratings and for maturities from 1 to 5 years

	1	2	3	4	5
AAA	1.10%	1.46%	2.39%	2.73%	3.29%
AA	1.12%	1.49%	2.52%	2.76%	3.33%
A	1.15%	1.52%	2.25%	2.94%	3.52%
BBB	1.26%	1.68%	2.37%	3.17%	3.65%
BB	1.73%	2.43%	3.24%	4.14%	4.89%
B	1.91%	2.66%	3.72%	4.82%	5.68%
CCC	5.29%	7.00%	8.64%	9.38%	9.98%

Hypothesis

In this example we consider a bank that invested into the previously structured EU CDS fund, composed by 55 European CDS. For each CDS the bank spent €1,000, for a total value of €55,000. Furthermore the bank is the buyer with respect to the reference entity and the seller respect to fund's underwriters. The overall return to the bank from this investment is 5% (each CDS gives a 5% return rate). Lastly if a CDS goes into default the estimated loss is €600 (provided a recovery rate RR of 0.4).

CM Model

First thing to do is estimating the 1-year maturity transition matrix that shows us probabilities of changing class, upgrading, or downgrading. The overall rating associated to the fund is BBB class and the 5% return gives back to the bank €2,750 yearly. The group of CDS considered has 1-year maturity, and spot rates have been computed taking into consideration the data related to the *Monthly Yield Curve* (Table 6.5).[6]
The different possible values of the fund depending on which will be the rating evolution are reported in Table 6.7.
In recent years spot rate curves have been decreasing, and this is reflected also in our CDS pricing, in fact we obtained a value really close to the nominal. On each CDS the bank invested constant shares, and now the fund's value has to be computed taking into account every possible future scenario. The migration table[7] in Table 6.6 predicts how probable is the in rating classes.
We consider now a portfolio of CDS with maturity 1 year and overall rating BBB. Based on Table 6.6, we are able to associate to our portfolio (rating BBB) the right

[6]The data regarding spot rates have been provided from IRS Agency. The IRS is a bureau of the Department of the Treasury and one of the world's most efficient tax administrators. The website is the following: "http://www.irs.gov/Retirement-Plans/Monthly-Yield-Curve-Tables".

[7]The 1-year and 5-year transition matrixes were derived from CreditMetrics paper issued by J.P. Morgan. Ratings are provided by the Agency Standard & Poor's issued in 2014.

Table 6.6 1 -year maturity migration table

	AAA	AA	A	BBB	BB	B	CCC	D	*cum*
AAA	85.90%	13.43%	0.46%	0.07%	0.085%	0.02%	0.04%	0.00%	100%
AA	0.66%	90.41%	8.10%	0.66%	0.07%	0.08%	0.01%	0.01%	100%
A	0.03%	1.92%	91.80%	5.64%	0.37%	0.15%	0.02%	0.07%	100%
BBB	0.01%	0.14%	4.63%	83.98%	5.81%	3.23%	1.94%	0.26	100%
BB	0.01%	0.03%	0.15%	5.71%	84.81%	7.70%	0.73%	0.84%	100%
B	0.00%	0.03%	0.11%	0.24%	6.13%	84.12%	4.96%	4.40%	100%
CCC	0.00%	0.00%	0.17%	0.26%	0.77%	15.69%	51.89%	31.22%	100%

Table 6.7 Fund value with $CreditMetrics^{TM}$ in the short maturity case

	V	*p*	*cump*
AAA	57,722.77	0.01%	0.01%
AA	57,719.54	0.14%	0.15%
A	57,718.73	4.63%	4.78%
BBB	57,715.78	83.98%	88.76%
BB	57,703.23	5.81%	94.57%
B	57,698.46	3.23%	97.80%
CCC	57,611.87	1.94%	99.74%
D	27,300	0.26%	100%

probability to change to a new rating class. Moreover, given the portfolio's values reported in Table 6.7, the expected value is then $\mu_V = \mathbb{E}[V] = 57,633.54$.
This way to recover the value of the portfolio at 99% confidence level, we use the linear interpolation method, yielding $V_{99\%} = 57,643.77$. The portfolio in the 99% probability will be worth €57,643.77 so the VaR in 99% of scenarios will be the difference between portfolio's expected value μ_V and this estimation, i.e., $VaR_{99\%} = \mu_V - V_{99\%} = 10.23$. Nevertheless this procedure excludes the standard deviation related to the portfolio. In order to include this element, we recover the total VaR at 99% confidence level which is the sum of $VaR_{99\%}$ and σ_{Total}, i.e., $TVaR_{99\%} = 5,736.55$, where

$$\sigma_{Total} = \sqrt{\sum_{j=1}^{8}(p_j \times V_j^2) - \mu_V^2} = 5,736.55$$

The total VaR at 99% confidence level which is called $TVaR_{99\%}$ is the sum of $VaR_{99\%}$ and σ_{Total}, i.e., $TVaR_{99\%} = 5,746.79$.

HYC Model

To compute the VaR derived from HYC model, we calculate the expected loss using the pdf found with the algorithm. The default probability needed is on the 1-year maturity line, i.e.,

$$EL_1 = pd_1 \times LGD_1 \times EAD_1 = 37.72 \tag{6.1}$$

The expected loss in equation (6.1) shows that for the existent PD, i.e., the default probability in the scenario with only one default, the capital needed to cover losses is €37.72. The same operation is replied for every possible scenario, which is 55 plus the cases where there is no default. Adding all EL available, the value produced would reflect the capital to retain in order to cover all expected losses, i.e.,

$$cEL_{55} = \sum_{i=0}^{55} EL_i = 12,763.73$$

We compute the losses at the 99% of confidence through a linear interpolation, yielding $cEL_{99\%} = 10,745.31$, which is the VaR at 99% level of confidence. VaR is composed by *expected loss* and *unexpected loss*; EL is considered to be as the mean of the cumulative loss distribution, while VaR has to cover both the expected and unexpected losses, but not the extreme losses. In the example at hand, the expected and unexpected losses are $EL_{mean} = 6,193.80$ and $UL_{99\%} = 4,551.51$, respectively. Therefore we have that $TVaR_{99\%} = EL_{mean} + UL_{99\%} = 10,745.31$. In this case our analysis showed that with HYC model the bank should have set aside more capital to cover the risk assuming a more conservative approach than that based on CM.

Finally we compare the HYC-VaR and the CM-VaR which depends on transition matrices and credit rating agencies' decisions. The transition matrix is a component which assumes a relevant role in CM where the computation of portfolio's value implies data in this matrix as weights. Capital requirements for CM with the 1-year maturity portfolio corresponds to 10.43% of the total portfolio value. On the other hand, HYC suggests to set aside 23.2% of the portfolio's value, seemingly more conservative than CM. CM here always underestimates VaR, thus leading to possible less conservative risk management policy. Furthermore this does not mean a constant under-evaluation behavior; we could detect also an overestimating event sometimes, due to the particular dependence structure implied in the transition matrix reported by the rating agencies.

6.3 Structural and Marginal Distortion in a Credit-Exposed Portfolio: a DHC Application

In this section we want to test the relevance of the dependence structure and of the distortion induced by the arrival policy of defaults linked to a portfolio composed by credit-exposed assets in such a way to avoid over/under valuation of the involved risk (see Bernardi and Romagnoli 4). We point out that to implement a protective strategy without drawing upon the risks' integration techniques, we would hedge the default risk with univariate credit derivatives, such as credit default swaps[8] (CDSs for short). Obviously such a strategy is not conditioned by the dependence structure of the multivariate portfolio because it is essentially univariate, its price being the algebraic sum of several univariate products. On the other hand, a multivariate hedging strategy, which could be implemented for a partial/total hedging by a portfolio of *nth to default* products, has a price that depends on the loss distribution of the portfolio. Therefore the choice of the best hedging policy, given an optimal criterion, such as minimizing a risk measure computed for a given risk aversion parameter, cannot necessarily leave out of consideration the integration of risks problem. With this in mind, a right valuation of the loss distribution becomes a pivotal ingredient in decision-making since a too conservative loss cdf, assigning a greater price to the *nth to default* products, would consequently result to be not optimal for the undervaluation of the healthy consequences of the dependence structure even though characterized by a more prudent behavior to risk.

We structure a portfolio of assets issued by some of the major banks of the European area that is exposed to the corresponding marginal credit risks and to the correlations among risks. All data are drawn from Thomson Reuters Datastream. We choose ten banks in the European area, and we construct a data set composed by the CDS prices (for five relevant maturities, i.e., from 1 to 5 years) on 15 July 2014, inspired by [4].

To implement the DHC procedure, we need to extrapolate from the market data, information concerning the marginals, i.e., the default probabilities and the dependence structure. We know that CDS market is very liquid and it competes with equity for obligor-specific information and it helps us to investigate the term structure of the default probability. In this case it is possible to *bootstrap* the default probabilities implicit into the prices in a manner similar to the one by which we usually back out the zero-coupon factor curve from swap rates. Therefore starting from CDSs' prices and by equating the protection and the premium legs with a recursive forward procedure (see Cherubini and Della Lunga 6), we got implicit default probabilities assuming a LGD[9] that we fixed at 0,6. The bootstrapped default probabilities, for the discounting factors implied by the EURIRS on 15 July 2014 for the different maturities, are listed in Table 6.8. These probabilities reflect the information concerning the banks' credit risk as a consequences of fluctuating market prices.

In this example we focus our attention to a midterm risk involved in the previous specified portfolio, whose marginal probabilities of default $\mathbf{p}$ are represented by the

---

[8]A CDS is a proper credit derivative product designed to transfer the credit exposure of fixed income products between parties, protecting the buyer against credit events that typically are defaults.

[9]LGD is an attribute of any exposure and represents the share of an asset that is lost when borrower defaults. LGD is facility-specific because such losses depend on several characteristic such as the presence of collateral and the degree of subordination. Nevertheless Basel II requirements prescribe fixed LGD ratios (from 0.45 to 0.75) for classes of unsecured exposures; our portfolio can be assumed to belong to. Obviously the chosen value is not restrictive for our procedure.

Table 6.8 Bootstrapped default probabilities

BANK-COUNTRY	MATURITY				
	1Y	2Y	3Y	4Y	5Y
BANCO SANTANDER-SPAIN	0.005413	0.015986	0.033225	0.059608	0.108234
LLOYDS BANK-UK	0.002928	0.009909	0.022231	0.040714	0.07954
BNP PARIBAS-FRANCE	0.003618	0.011159	0.02558	0.04664	0.08628
DEUTSCHE BANK-GERMANY	0.00456	0.014153	0.02959	0.05339	0.09729
BANCO ESPIRITO SANTO-PORTUGAL	0.127638	0.1200562	0.23547	0.270610	0.277564
INTESA SANPAOLO SPA-ITALY	0.005478	0.016884	0.035725	0.06121	0.106955
KBC BANK-BELGIUM	0.003168	0.010271	0.022909	0.056496	0.12121
NORDEA BANK-SWEDEN	0.001588	0.004346	0.009212	0.026023	0.060118
CREDIT SUISSE GROUP-SWITZERLAND	0.003497	0.009987	0.022521	0.039572	0.070509
DNB BANK-NORWAY	0.000795	0.008064	0.023046	0.049883	0.068003

last column of Table 6.8, i.e., the bootstrapped probabilities from CDS prices with maturity of 5 years.

We now consider the dependence structure that plays an important role both at the first level and at the second one, representing the dependence among the ordered couples of variables. For this purpose, we choose to work with the Archimedean class of copula functions which allows to reduce the study of a multivariate copula to a single univariate function. In this class we select three families of Archimedean copulas, i.e., Clayton, Gumbel, and Frank, identifying a subclass that comprehensively covers all possible tail dependency assumption. Given a decreasing and convex function $\phi : (0, 1] \to [0, +\infty)$ such that $\phi(1) = 0$, then the function

$$C^{(\phi)}(\mathbf{u}) = \phi^{-1}\left(\sum_{i=1}^{n} \phi(u_i)\right), \qquad \forall u_i \in (0, 1], \forall i = 1, \dots, n$$

where $\mathbf{u}$ stands for the vector of margins is a n-dimensional Archimedean copula with generator ϕ. Different choices of generator define several important families of Archimedean copulas. It is important to note that this class of copulas is widely used, especially in multivariate applications, since the dependence structure among variables can be defined only by means of a parameter α, functionally linked to a correlation measure as Kendall's τ coefficient.

To calibrate the dependence structure, we applied the Genest and Rivest's [1993] procedure that is detailed in Frees and Valdez [7] based on the time series of CDSs' prices with maturity 5 years from 17 June 2008 to 15 July 2014. More precisely if for simplicity we consider the bivariate case, given a set of observations $(X, Y)_j$, $j = 1, \dots, n$, our aim is to identify the form of the generator ϕ specifying the Archimedean copula $C^{(\phi)}$ associated to the cumulated probability distribution function $F_{(X,Y)_j}$. Denoting $Z_j = F^j_{X,Y}$, Genest and Rivest [9] prove that the probability distribution function of Z_j is linked to the generator of an Archimedean copula as follows:

$$F_{Z_j}(t) = t - \frac{\phi(t)}{\phi'(t)},$$

and suggest to identify ϕ through the following steps: (i) estimate the Kendall's correlation coefficient as $\tau_n = \binom{n}{2}^{-1} \sum_{i<j} \text{sign}[(X_i - X_j)(Y_i - Y_j)] and$ (ii) construct a nonparametric and a parametric estimate of $F_{Z_j}(t)$. To address the last one, we observe that given the estimate τ_n, we can compute an estimate dependence parameter α_n which can be used to estimate the generator $\phi_n(t)$ and finally $F^{(\phi_n)}_{Z_j}(t)$. The parametric estimation must be repeated for different choices of the generator and then compared with the nonparametric estimate $F^{(n)}_{Z_j}(t)$. Finally, a goodness of fit consideration allows to select the "closest" family of copula functions. The detailed technique applied here uses a multistage procedure in the same spirit of

Okhrin et al. [12]; first of all we estimate the parameter at the lowest level of the hierarchy, and at further stages the next level copula's parameter is estimated assuming that the copula parameters at lower levels are known. This way we implement a recursive estimation at each level of the hierarchy depending on the copula parameters at lower levels; at each level we recover a mean-squared error (MSE for short) that we call partial MSE (PMSE for short) and whose mean among all levels comes into what we call total MSE (TMSE for short). We estimate a dependence structure at the first level represented by a Kendall tau, $\tau_1 = 0.428$ and a second-level vector of decreasingly ordered dependence parameters representing the dependence among pairs of variables,[10] i.e.,

$$\boldsymbol{\tau_2} = \left[0.835\ 0.707\ 0.645\ 0.614\ 0.614\right].$$

We consider the following Archimedean dependence structure:

1. One-level dependence structure, represented by a copula function with dependence parameter τ_1
2. Partially exchangeable hierarchical structure, with two dependence levels represented by τ_1 and $\boldsymbol{\tau_2}$
3. Not exchangeable partially nested hierarchical structure, whose nesting levels are given by $\boldsymbol{\tau_2}$

We can observe that the dependence structure represents the relationships between the defaults; subsequently it has an impacts on their arrival through the arrival matrix **A**. This matrix defines the distorted combinatorial distributions that are then aggregated with the same assumed dependence structure. Indeed it is evident that the impact of the dependence structure is twofold since it enters into the determination of the arrival matrix and finally acts onto the aggregating step. The novelty of the DHC is precisely to introduce the arrival policy as a determinant of the loss distribution through a distortion of the combinatorial distribution. If we assume independence, i.e., we assume that the arrival of defaults are independent of the dependence structure of variables that impacts only the aggregation, the arrival matrix will be distorted only by the marginals probabilities that attract the defaults with different forces. In fact the marginal probabilities of defaults contribute in determining the arrival matrix along with the dependence structure; to identify the contribution of the dependence structure, we consider the case of homogeneous marginals equal to the centroid of the set of the default's probabilities characterized by the vector **p**. In this case the force of attraction will be the same for all the variables, and then the distortion will depend only on the assumed dependence structure; if we work with the independence assumption, the arrival matrix will converge to the not distorted combinatorial distribution matrix that defines the CHC

[10]The couples of variables are ordered such to have a decreasing dependence parameters. The couples are Banco Santander-Intesa San Paolo, Deutsche bank-Nordea bank, Credit Suisse Group-DNB bank, KBC-Banco Espirito Santo, and Lloyds bank-BNP Paribas.

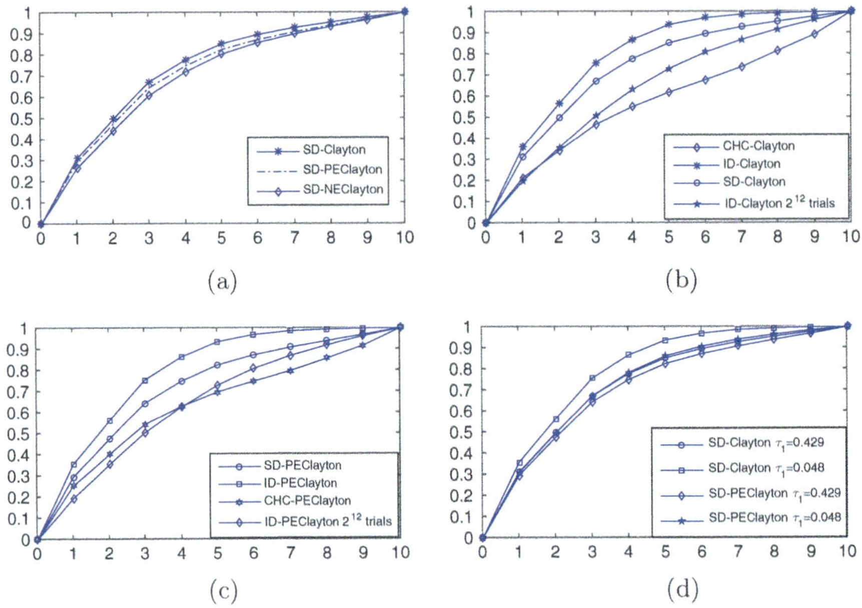

Fig. 6.3 Structural distortion on the loss cdf for several dependence structure. (**a**) Clayton, PEClayton, and NEClayton. (**b**) Clayton structure vs CHC-Clayton and ID-Clayton for different number of trials. (**c**) PEClayton structure vs CHC-PEClayton and ID-PEClayton for different number of trials. (**d**) Clayton and PEClayton for different τ_1

approach for a sufficient large number of trials (see Bernardi and Romagnoli 2). In the base scenario, we decide to compute an arrival matrix of dimension $d \times n$, where $n = 10$ is the number of assets into the portfolio, while $d = 2^{10}$ corresponds to the number of combinatorial distributions of defaults. Moreover we choose to work with a number of trials $N = 2^{11}$.

In the following we model, the dependence structure through the best performing Archimedean copula, i.e., the Clayton copula, and then we consider the listed cases of aggregation, i.e., the not hierarchical Clayton, and the partially/not exchangeable version of the hierarchical Clayton (PEClayton and NEClayton, respectively)[11] for the dependence parameters computed starting from the estimated Kendall τ_1 and $\boldsymbol{\tau_2}$.

In Fig. 6.3a we plot the cumulative probability distribution functions of the number of defaults computed through the DHC approach where we assume homogeneous marginal default probabilities. Here we can see the impact of the dependence structure on the arrival policy that we call *structural distortion* (SD),

[11]The sampling from the hierarchical Archimedean structures has been done with the algorithm proposed in Savu and Trede [14]. This algorithm, based on the conditional inversion method, is computationally intensive especially in high dimensional problems. For a better performing sampling, see Hofert and Mächler [10].

for different dependence structures. We may observe that the structures with at least two-level dependence are in general a little bit less conservative than those at one level, this showing that the hierarchy doesn't impact much in this example, where we have high dependence still at the first level. We point out that a less conservative loss cdf corresponds to a less protective attitude to risk, usually implying a more efficient management in term of retained monetary reserves but also undertaking the undervaluation of risk by effect of a too optimistic assessment of the loss distribution.

In order to evaluate the impact of the dependence structure on the distortion of the loss cdf, we make a comparison of the loss distribution computed through the DHC approach with homogeneous marginals, where the arrivals are assumed to be independent of the dependence structure among variables (ID for short) and the CHC loss distribution. This last case does not generate any distortion since the arrival policies are not considered in the model, as discussed in Chap. 2. In Fig. 6.3b, c, we make the comparison considering the Clayton and the PEClayton structure, respectively. The behavior of the loss cdfs shows that in general the result of a structural distortion is a more conservative curve than the CHC model where no distortion acts. Therefore we can appreciate the convergence of the ID case to the CHC one, for a large number of trials.

Moreover the structural distortion depends on the dependence parameters. To assess this sensitivity, we consider a scenario characterized by a lower first-level Kendall $\tau_1 = 0.048$ that is very near to the independence case, to be compared with the base scenario. In Fig. 6.3d we plot the loss cdfs corresponding to the homogeneous one-level and partially exchangeable Clayton structure into the two scenarios. The behavior of the curves shows that the effect of a lower first-level Kendall τ_1 is a more conservative loss cdf and the hierarchy seems to weaken this effect, confirming the evidence of the previous analysis. This evidence highlights the positive effect of a systemic factor in the evaluation of the considered portfolio's risk.

As pointed out before, the impact of the arrival policy on the loss distribution depends also on the marginal default probabilities that represent the force of attraction for the default event. In Fig. 6.4a we plot the loss cdf for our not homogeneous variables, where the dependence structure does not impact the arrival policy. This analysis allows us to disentangle the effect of the inhomogeneous marginals on the arrival policy that represents a partial distortion called *marginal distortion*. The different dependence structures considered here act only in the aggregation step since we do not consider here any *structural distortion*. The qualitative behavior of the curves shows that the marginal distortion does not have a strong impact, since the dependence structure enters only in the aggregation step but does not modify the arrival matrix.

On the other hand, the marginal distortion depends on the degree of dispersion of the marginal default probabilities around the centroid represented by an ideal variable whose default probability is the nearest, given a distance measure, to the marginals of the portfolio. To analyze the sensitivity of the marginal distortion with respect to the dispersion of marginals, we consider a maximum scattered scenario

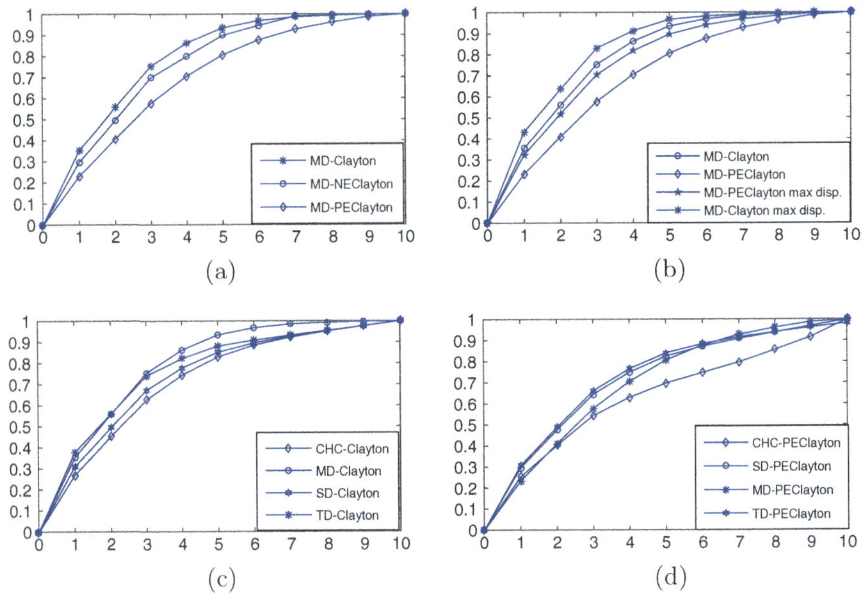

Fig. 6.4 Partial and total distortion on the loss cdf for several dependence structure. (**a**) Marginal distortion for Clayton, PEClayton, and NEClayton. (**b**) Marginal distortion for Clayton and PEClayton for different degree of marginals' dispersion: base vs maximum dispersion scenario. (**c**) Total and partial distortions for Clayton vs the corresponding not distorted cdf. (**d**) Total and partial distortions for PEClayton structure vs the corresponding not distorted cdf

where the marginals are concentrated around the extreme default probabilities that are 0.060118 and 0.277564 in our example, i.e., the min/max default probabilities for maturity 5Y as reported in Table 6.8 but where the centroid is the same as that of the base scenario. If the centroid does not change, we can assume that the structural distortion will not be modified, while the marginal distortion will be. We compare the loss cdfs in picture Fig. 6.4b; we can observe that the impact of the greater dispersion is more evident in the hierarchical Archimedean dependence structure since at the second level the dispersion interacts with a stronger degree of dependence. On the other hand, the effect of a greater dispersion on the marginals is a more conservative curve than the one of the corresponding base scenario, showing that the effect of the marginals' dispersion prevails on that of the dependence structure acting here only in the aggregation phase.

In Fig. 6.4c, d, we compare the partially distorted loss cdfs by effect of the structural or the marginal distortion, with the totally distorted and the not distorted ones, for several dependence structures. Here it is possible to evaluate the impact of the distortion and to disentangle the effect of the structural and the marginal impact. The one-level Clayton structure shows that the two partial distortions interact generating a total distortion curve that is less conservative than the marginally distorted cdfs. On the other hand, the partially exchangeable hierarchical Archimedean structure produces a total distorted cdf that generally dominates the partially distorted ones showing that the partial distortions go in the same direction summing each other without any compensation; in this empirical example, the hierarchy acts amplifying the effect of the lower tail dependence. The not distorted curve is always less conservative than the totally and partially distorted curve showing the importance of including the arrival matrix in the model in order to avoid an undervaluation of risk.

6.4 A Bayesian Analysis of the DHC Model

In this analysis, referred to Gaudenzi [8], we compare the market evaluation of financial products through a Bayesian model. In particular we'll consider a portfolio which replicates a quoted index, and we'll compare their loss functions through the analysis of the errors. This is a Bayesian analysis because it is based on the feature of the posterior distribution which allows us to implement hypothesis tests.

A portfolio is considered composed by iTraxx Europe Senior Financials Series 13's constituents since iTraxx indices are used as proxies for default risk. The iTraxx Europe Senior Financials is a sub-group of iTraxx Europe index, and it is composed by 25 equally weighted sub-investment grade entities which are the most liquid in Europe. The composition, i.e., 5-year CDS portfolio, is updated on semiannual basis. This index is traded like a CDS on a single firm, and if a constituent default, it is removed from the index, and the nominal value of the contract declines by 2%. Data was extracted from Thomson Reuters Data Streams for the period between 22 March 2010 and 30 September 2010. To recover the marginal distributions, i.e., the default probabilities of the constituents, and to calibrate the copula function representing the dependence structure of the portfolio, we'll use the prices of the CDS products concerning the constituents themselves along with the prices of the CDSs for different maturities (from 1 year to 5 years) and concerning the period from 21 September 2009 and 30 September 2009.
The representation of the multivariate dependence structure is performed in two steps: first we estimated the marginal distribution of each assets separately from the dependence structure, and then we assessed the dependence structure among them. From the CDSs prices, we recovered the marginal default's probabilities by the bootstrapping method where we assumed a loss given default of 60% as suggested by Markit for the senior credits of the financial sector. In Fig. 6.5 we report the default probabilities for maturity 5 years of the constituents on 9 September 2010. For the sake of simplicity, we decided to cluster in six sub-portfolios according to the marginal probabilities using the K-means algorithm. In this case the complexity of the clusters is given by the following vector:

$$[4 \quad 5 \quad 5 \quad 4 \quad 1 \quad 6].$$

As we can observe the singleton is represented by the Banco Espirito Santo which shows a default probability greater than 30%.

Fig. 6.5 The set of default probabilities of the iTraxx Europe Financial Series 13's constituents

Aegon	13, 25%
Allianz	5, 78%
Assicurazioni Generali	10, 87%
Aviva	9, 32%
Axa	9, 34%
Monte dei Paschi di Siena	16, 60%
Banco Bilbao Vizcaya Argentaria	16, 98%
Banco Espirito Santo	32, 54%
Banco Santander	11, 09%
Bank of Scotland	13, 78%
Barclays Bank	10, 09%
BNP Paribas	8, 54%
Commerbank	8, 62%
Credit Agricole	10, 82%
Credit Suisse Group	8, 64%
Deutsche Bank	5, 34%
Hannover Rueckversicherungs	7, 76%
Intesa Sanpaolo	10, 75%
Muenchener Rueckversicherungs	5, 06%
Societe Generale	10, 13%
Swiss Reinsurance	10, 53%
The Royal Bank of Scotland	13, 89%
Ubs	8, 73%
Unicredit	12, 51%
Zurich Insurance Company	6, 77%

6.4.1 Multivariate Dependence Calibration

In order to define the multivariate dependence, we proceed with an econometric filtration based on the log returns of the sub-portfolios. The stationarity of the time series is evaluated and confirmed through the *Augmented Dickey-Fuller test*. Then we analyze the possible autocorrelation of the residuals through the *Ljung-Box Q-test*: the returns of the sub-portfolios 3, 4, 5, and 6 are strongly autocorrelated, while portfolios 1 and 2 accept the null hypothesis of no autocorrelation with a significance level greater than 10% for the first eight lags. The results of such test indicate that there is not enough evidence to reject the null hypothesis that the returns are not autocorrelated. However running the same test on the squared returns, it suggests that they are not independent, meaning that there is ARCH effect in the returns of the time series. To be sure of that, we test for return heteroscedasticity through the *archtest* which confirms the presence of ARCH effects in all sub-portfolios with exception for some lags in portfolios 3, 5, and 6. Moreover the degree of persistence in variance implies that a GARCH model may be well suited for our data set. Then finally to remove the autocorrelation from the original series, we apply the ARMA filter, while the ARCH model is implied to model the returns' volatility. This way the ARMA-GARCH model works as a filter which provides serially independent innovations which are then standardized and used in copula's calibration. For a large number of sub-portfolios, a model ARMA(1,1)-GARCH(1,1) is adopted, while for other sub-portfolio, the autocorrelation of residuals was neutralized only after two lags. Only in the first sub-portfolio, the autocorrelation was removed with an ARMA(0,0) model. The adequacy of the estimated models is assessed graphically by drawing the sample autocorrelation function (ACF) plots of the standardized residuals which are reported in Figs. 6.6. As we can observe, they don't reveal either serial correlation or dependence at least for ten lags. We are now ready to estimate the copula's parameters and determine its structure using the recovered standardized residuals. We choose to perform the estimation using a nonparametric model, i.e., through the empirical distribution functions, using the *HAC* and *Nacopula* packages. The method used in the copula estimate is the *recursive ML* since it holds more precise parameters estimate as suggested in Okhrin and Ristig [13]. Moreover we select to work in the Hierarchical Archimedean family (with six classes), and finally we implement a GOF test which reveals that the optimal choice is the Clayton copula. As we can see in Fig. 6.7, the best structure which represents our data set is a fully nested Hierarchical Archimedean copula. Table 6.9 reports the Kendall's τ values among sub-portfolios, while Fig. 6.8 visualizes 500 sampled random vectors from the estimated copula and represents them with a scatter plot matrix. From this figure it is possible to note the lower dependence between sub-portfolios in the higher part of the three; sub-portfolios 2, 3, and 4 have a higher Kendall's τ, and their scatter plot is more linear and less dispersed than the others, while sub-portfolio 5 shows the lowest Kendall's τ and a scatter plot representing more independent and equally distributed simulation than the others. Moreover we can appreciate in Fig. 6.8 the lower tail dependence of the sub-portfolios and the absence of upper tail dependence due to the optimality of the copula.

6.4.2 The Loss Function: Index Versus Replicating Portfolio

To recover the arrival matrix which is an important input of the DHC model, we need to sample from the calibrated nested copula.[12] As we know this is a stochastic matrix of 0–1 entries representing the most probable scenarios of defaults' arrival, given the marginal default's probabilities of groups and the dependence structure. Every group is assumed to be homogeneous for that concerns the features which impact on the default's event, i.e., inside the same group they go into default at the same time.

[12]This is done by the *rnacopula* function in R.

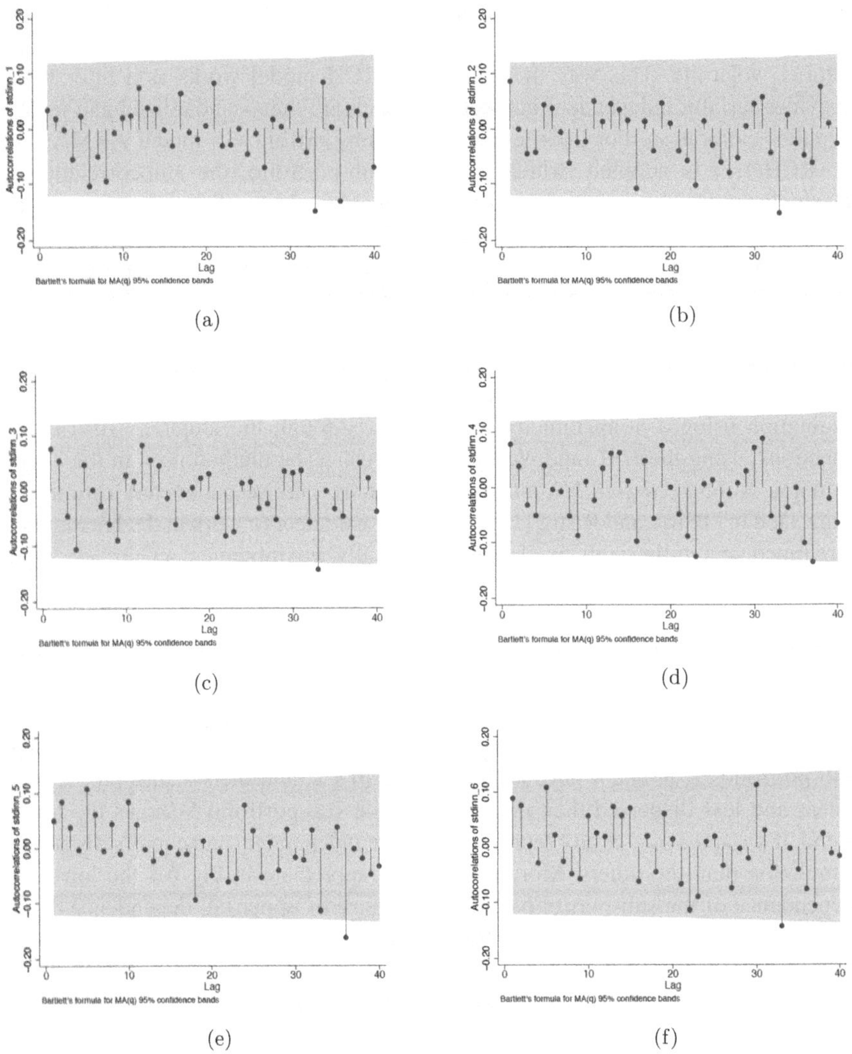

Fig. 6.6 ACF plots of sub-portfolios which replicate the iTraxx Europe Financial Series 13. (**a**) ACF plot related to sub-portfolio 1. (**b**) ACF plot related to sub-portfolio 2. (**c**) ACF plot related to sub-portfolio 3. (**d**) ACF plot related to sub-portfolio 4. (**e**) ACF plot related to sub-portfolio 5. (**f**) ACF plot related to sub-portfolio 6

This way we are able to simulate an arrival matrix and to recover the corresponding loss function of the portfolio by the DHC algorithm.

On the other hand, the loss function of the iTraxx Europe Financial Series 13 index has been recovered based on the default probability of the index recovered by its market price. To do this we consider the index as a *cds* product, and we recover a default probability of 11.19%. Then finally we are able to recover the loss function

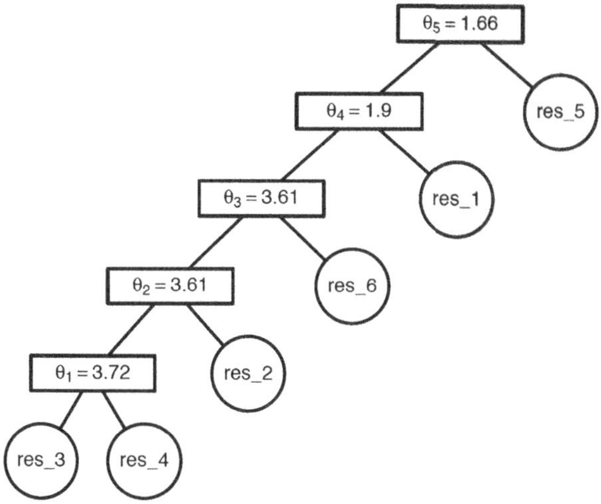

Fig. 6.7 Optimal dependence structure explaining the portfolio which replicates the iTraxx Europe Financial Series 13

Table 6.9 Kendall's τ values of the sub-portfolios which replicate the iTraxx Europe Financial Series 13

Sub-Portfolio	Sub1	Sub2	Sub3	Sub4	Sub5	Sub6
Sub1	1	0.5773	0.5926	0.6110	0.4747	0.5830
Sub2	0.5773	1	0.7439	0.7209	0.5892	0.7535
Sub3	0.5926	0.7439	1	0.7546	0.5899	0.7406
Sub4	0.6110	0.7209	0.7546	1	0.5722	0.7197
Sub5	0.4747	0.5892	0.5899	0.5722	1	0.5733
Sub6	0.5830	0.7535	0.7406	0.7197	0.5733	1

of the index by a simulated frequentist approach where all groups are homogeneous in term of default's probability. In Fig. 6.9 we compare the loss function of the index and that of the replicating portfolio. We observe that the loss function of the index lies above that of the replicating portfolio in all the domain of the function, meaning that the market attributes a greater probability to scenarios represented by a small number of defaults (till 8 defaults) and a lower probability to scenarios involving a big number of default's events.

6.4.3 A Bayesian Analysis on the Residuals

The aim is to study by a Bayesian inference method the residuals corresponding to the distance between the loss functions of the index and that of the replicating portfolio. This way by a Bayesian regression on the residuals, we recover the

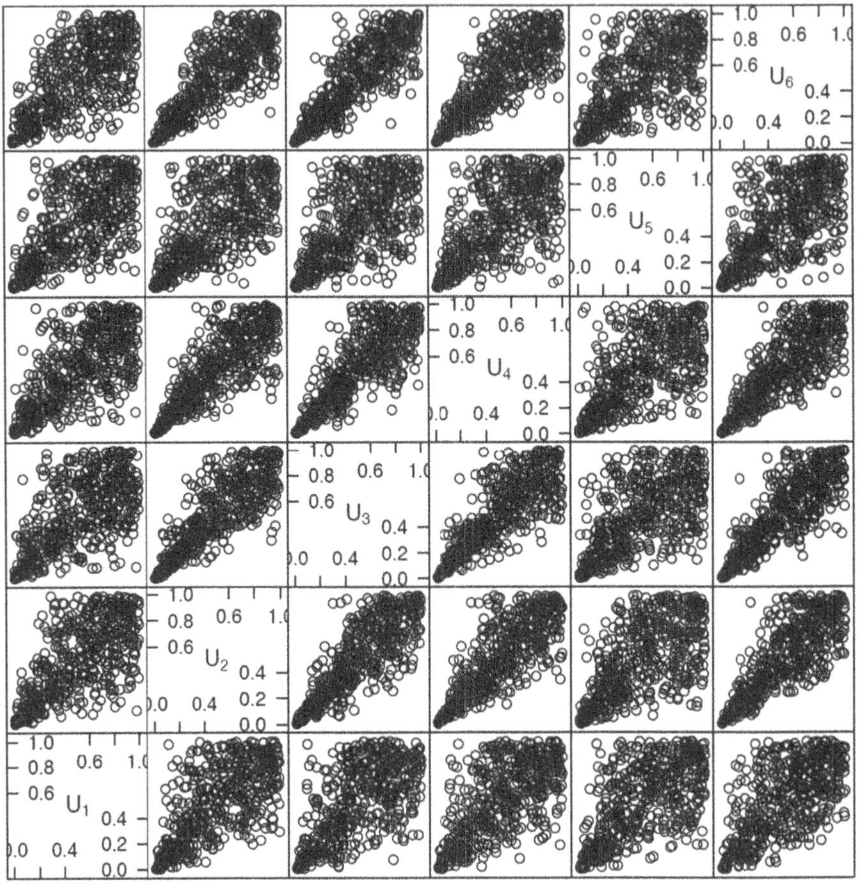

Fig. 6.8 Scatter plot of 500 sampled random vectors from the estimated copula of the sub-portfolios which replicate the iTraxx Europe Financial Series 13

posterior distribution of the sample mean of the residuals. In Bayesian model we need to specify a likelihood, i.e., a conditional distribution of data and we need to choose an a prior distribution based on our information. Table 6.10 reports the descriptive statistics of the residuals where the resulting mean is negative, as expected given the dominance of the index's loss function. Moreover from the reported values of Skewness and Kurtosis, we choose a normal distribution of the residuals and a non-informative prior, i.e., a uniform distribution.

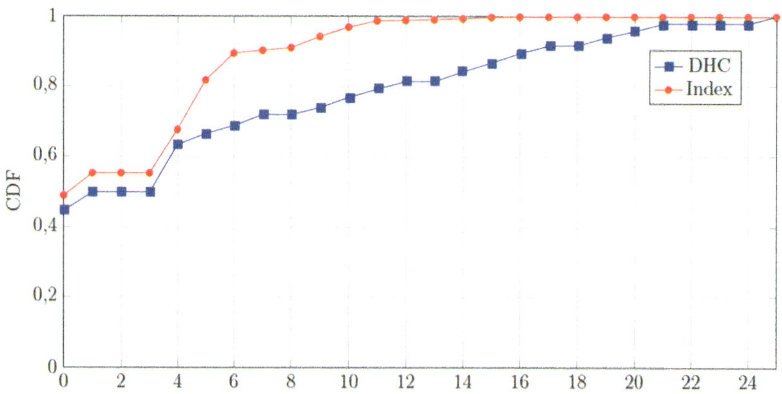

Fig. 6.9 Loss distribution of the iTraxx Europe Financial Series 13 and of the replicating portfolio

Table 6.10 Descriptive statistics of the residuals

Observation	Mean	Standard deviation	Skewness	Kurtosis
520	−0.1162783	0.0779901	−0.2511596	1.785397

Table 6.11 Bayesian analysis: estimated values of the parameters

Mean	Standard deviation	MCSE	Median	95% confidence interval
−0.1161411	0.0034415	0.0001	−0.1161864	(−0.12293;−0.10910)

Table 6.11 reports the estimated values of the parameters. The Bayesian analysis gives a strongly negative mean, and based on the posterior, we have that with a confidence of 95% the unknown mean lies in interval relatively far from zero.

Figure 6.10 shows the posterior distribution of the mean of residuals which confirms the previous assertion. Then finally we can say that the index attributes higher probability on the left queue and lower probability on the right queue than the replicating portfolio, meaning that the market overvalued the index assuming a lower risk than the replicating portfolio and consequently the protection sellers didn't save enough capital to cover the expected losses coming from their position on the index. Moreover regarding the right queue of the loss distribution, we stress that the DHC distribution takes into account the systemic risk and contagion effects among constituents coming from the recovered arrival matrix, while the market seems to undervalue those risks probably due to the historical and structural features of the financial market in the period of analysis, but we cannot discard the possibility of a wrong market evaluation of the risk.

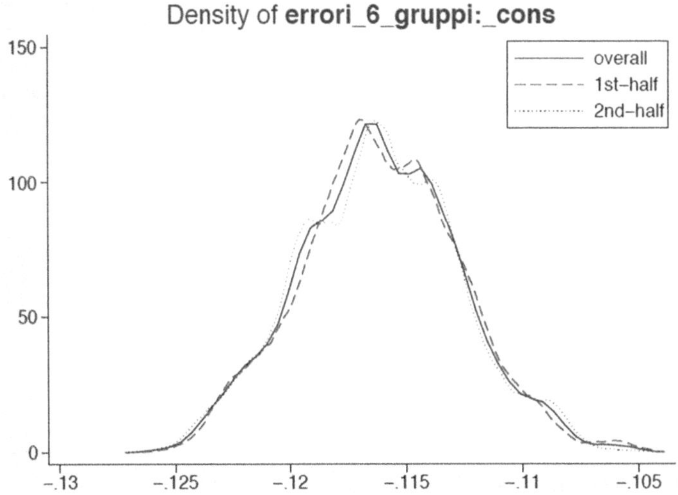

Fig. 6.10 Posterior distribution of the mean of residuals

References

1. Bernardi, E., & Romagnoli, S. (2011). Computing the Volume of an high-dimensional semi-unsupervised Hierarchical copula. *International Journal of Computer Mathematics, 88*(12), 2591–2607.
2. Bernardi, E., & Romagnoli, S. (2013). A Clusterized copula-based probability distribution of a counting variable for high-dimensional problems. *The Journal of Credit Risk, 9*(2), 3–26.
3. Bernardi, E., Falangi, F., & Romagnoli, S. (2015b). A hierarchical copula-based world-wide valuation of sovereign risk. *Insurance Mathematics & Economics, 61*, 155–169.
4. Bernardi, E., & Romagnoli, S. (2016). Distorted copula-based probability distribution of a counting hierarchical variable: A credit risk application. *International Journal of Information Technology & Decision Making, 15*, 285–310.
5. Bernardi, E., Romagnoli, S., & Doti, M. (2019). The impact of the dependence structure in risk management: A focus on credit-risk. *International Journal of General Systems, 48*(4), 335–361.
6. Cherubini, U., & Della Lunga, G. (2007). *Structured Finance*. Chichester: Wiley Finance Series.
7. Frees, E. W., & Valdez, E. (1998). Understanding relationship using copulas. *North American Actuarial Journal, 2*, 1–25.
8. Gaudenzi, R. (2016). Loss Function di un portafoglio multivariato: un approccio Bayesiano. Master Thesis, University of Bologna.
9. Genest, C., & Rivest, L. (1993). Statistical inference procedures for bivariate Archimedean copulas. *Journal of the American Statistical Association, 88*, 1034–1043.
10. Hofert, M., & Mächler, M. (2011). Nested Archimedean copulas meet R: The nacopula Package. *Journal of Statistical Software*, in press.
11. Kohonen, T. (1982). Self-Organized formation of topologically correct feature maps. *Biological Cybernetics, 43*, 59–69.
12. Okhrin, O., Okhrin, Y., & Schmid, W. (2013). Properties of hierarchical Archimedean copulas. *Statistics and Risk Modeling, 30*(1), 21–53.

13. Okhrin, O., & Ristig, A. (2014). Hierarchical Archimedean copulae: The HAC package. *Journal of Statistical Software, 58*(4), 1–20.
14. Savu, C., & Trede, M. (2010). Hierarchical Archimedean copulas. *Quantitative Finance, 10*(3), 295–304.
15. Schönbucher, P. (2003). *Credit Derivatives Pricing Models: Model, Pricing and Implementation*. Chichester: Wiley.
16. Wang, L., Bo, L., & Jiao, L. (2006). A modified K-means clustering with a density-sensitive distance metric. In *Rough Sets and Knowledge Technology*, Lecture Notes in Computer Science, vol. 4062, pp. 544–551. Springer

References

1. Amsler, C., Prokhorov, A., & Schmidt, P. (2010). Using Copulas to Model Time Dependence in Stochastic Frontier Models. working paper.
2. Ang, A., & Chen, J. (2002). Asymmetric correlations of equity portfolios. *Journal of Financial Economics, 63*(3), 443–494.
3. Asharaf, S., Narasimha Murty, M., & Shevade, S. K. (2006). Rough set based incremental clustering of interval data. *Pattern Recognition Letters, 27*, 515–519.
4. Azzalini, A., & Capitanio, A. (1999). Statistical applications of the multivariate skew normal distribution. *Journal of the Royal Statistical Society: Series B (Statistical Methodology), 61*(3), 579–602.
5. Beare, B. (2009). *Copulas and temporal dependence*. UC San Diego: Department of Economics.
6. Berrada, T., Dupuis, D. J., Jacquier, E., Papageorgiou, N., & Rémillard, B. (2006). Credit migration and derivatives pricing using copulas. *The Journal of Computational Finance, 10*, 43–68.
7. Bielecki, T., Jakubowski, J., & Nieweglowski, M. (2010). Dynamic modeling of dependence in finance via copulae between stochastic processes. In P. Jaworski, F. Durante, W. Haerdle, & T. Rychlik (Eds.). *Copula Theory and its Applications*. Lecture Notes in Statistics-Proceedings. Berlin/Heidelberg: Springer, pp. 33–76.
8. Bielecki, T., Jakubowski, J., Vidozzi, A., & Vidozzi, L. (2008a). Study of dependence for some stochastic processes. *Stochastic Analysis and Applications, 26*, 903–924.
9. Bielecki, T., Vidozzi, A., & Vidozzi, L. (2008). Multivariate Markov processes with given markovian margins and Markov copulae. working paper, Illinois Institute of Technology.
10. Billingsley, P. (1986). *Probability and Measure*, 2nd edn. New York: Wiley.
11. Bingham, N. H., Goldie, C. M., & Teugels, J. L. (1987). *Regular Variation*. Cambridge: Cambridge University Press.
12. Black, F., & Cox, J. C. (1976). Valuing corporate securities: Some effects of bond indenture provisions. *Journal of Finance, 31*, 351–367.
13. Black, F., & Scholes, M. (1973). The pricing of options and corporate liabilities. *Journal of Political Economy, 81*, 637–654.
14. Block, H. W., & Ting, M. L. (1981). Some concepts of multivariate dependence. *Communications in Statistics A - Theory Methods, 10*, 749–762.
15. Block, H., Savits, T., & Shaked, M. (1982). Some concepts of negative dependence. *Annals of Probability, 10*, 765–772.
16. Blomqvist, N. (1950). On a measure of dependence between two random variables. *Annals of Mathematical Statistics, 21*, 593–600.

E. Bernardi, S. Romagnoli, *Counting Statistics for Dependent Random Events*,
https://doi.org/10.1007/978-3-030-64250-1

17. Boero, G., Silvapulle, P., & Tursunalieva, A. (2010). Modelling the bivariate dependence structure of exchange rates before and after the introduction of the euro: A semi-parametric approach. *International Journal of Finance and Economics*, in press, available online.
18. Bouyè, E., & Salmon, M. (2009). Dynamic copula quantile regressions and tail area dynamic dependence in forex markets. *The European Journal of Finance, 15*(78), 721–750.
19. Bouyè, E., Durrleman, V., Nikeghbali, A., Riboulet, G., & Roncalli, T. (2000). Copulas for finance – A reading guide and some applications. Groupe de Recherche Opèrationelle, Crédit Lyonnais working paper.
20. Bradley, R. C. (2007). *Introduction to Strong Mixing Conditions*. Heber City: Kendrick Press.
21. Breeden, D. T., & Litzenberger, R. H. (1978). Prices of state-contingent claims implicit in option prices. *Journal of Business, 51*, 621–51.
22. Brèmaud, P., & Yor, M. (1978). Changes of filtrations and of probability measures. *Z. Wahrscheinl, 45*, 269–295.
23. Brindley, E. C., & Thompson, Jr W. A. (1972). "Dependence and aging aspects of multivariate survival". *Journal of the American Statistical Association, 67*, 822–830.
24. Buchinsky, M. (1998). Recent advances in quantile regression models: a practical guideline for empirical research. *Journal of Human Resources, 33*, 1, 88–126.
25. Burtschell, X., Gregory, J., & Laurent, J. P. (2007). Beyond the Gaussian Copula: Stochastic and local correlation. *Journal of Credit Risk, 3*(1), 31–62.
26. Cambanis, S., Huang, S., & Simons, G. (1981). On the theory of elliptically contoured distributions. *Journal of Multivariate Analysis, 11*, 368–385.
27. Cameron, A. C., Li, T., Trivedi, P. K., & Zimmer, D. M. (2004). Modelling the differences in counted outcomes using bivariate copula models with application to mismeasured counts. *Econometrics Journal, 7*, 566–584.
28. Carr, P., & Wu, L. (2004). Time changed Lèvy processes and option pricing. *Journal of Financial Economics, 17*, 113–141.
29. Cazzulani, L., Meneguzzo, D., & Vecchiato, W. (2001). Copulas: A statistical perspective with empirical applications. IntesaBCI Risk Management Department, *working paper*.
30. CEIOPS (2010) QIS 5 technical specifications: Technical report. Committee of European Insurance and Occupational Pensions Supervisors.
31. Chan, K., & Tong, H. (2001). *Chaos: A Statistical Perspective*. New York: Springer.
32. Chen, X., & Fan, Y. (2006). Estimation of copula-based semiparametric time series models. *Journal of Econometrics, 130*, 307–335.
33. Chen, X., Koenker, R., & Xiao, Z. (2009). Copula-based nonlinear quantile autoregression. *Econometrics Journal, 12*, S50–S67.
34. Chen, X., Wu, W. B., & Yi, Y. (2009). Efficient estimation of copula-based semiparametric Markov models. *Annals of Statistics, 37*(6B), 4214–4253.
35. Cherubini, U., & Luciano, E. (2001). Value at Risk trade-off and capital allocation with copulas. *Economic Notes, 30*(2), 235–56.
36. Cherubini, U., & Luciano, E. (2002). Bivariate option pricing with copulas. *Applied Mathematical Finance, 8*, 69–85.
37. Cherubini, U., & Romagnoli, S. (2010). The dependence structure of running maxima and minima: Results and option pricing applications. *Mathematical Finance, 20*(1), 35–58.
38. Cherubini, U., Luciano, E., & Vecchiato, W. (2004). *Copula Methods in Finance*. Wiley: London. (John Wiley Series in Finance).
39. Cherubini, U., Mulinacci, S., & Romagnoli, S. (2008). A copula-based model of the term structure of CDO tranches. forthcoming in Applied Quantitative Finance 2nd edn. Haerdle, Hautsch and Overbeck, Springer.
40. Cherubini, U., Gobbi, F., & Mulinacci, S. (2010a). Semi parametric estimation and simulation of actively managed portfolios. working paper.
41. Cherubini, U., Gobbi, F., Mulinacci, S., & Romagnoli S. (2010b). On the Term Structure of Multivariate Equity Derivatives. working paper.

42. Cherubini, U., Gobbi, F., Mulinacci, S., & Romagnoli, S. (2010c). A copula-based model for spatial and temporal dependence of equity markets. In F. Durante, W. Härdle, P. Jaworski, & T. Rychlik (Eds.). *Workshop on Copula Theory and its Applications*. Lecture Notes in Statistics-Proceedings. Berlin/Heidelberg: Springer.
43. Cherubini, U., Mulinacci, S., & Romagnoli, S. (2011). A copula-based model of speculative price dynamics in discrete time. forthcoming in Journal of Multivariate Analysis.
44. Cherubini, U., Mulinacci, S., & Romagnoli, S. (2011). On the distribution of the (un)bounded sum of random variables. *Insurance Mathematics and Economics, 48*(1), 56–63.
45. Clark, P. (1973). A subordinated stochastic process with finite variance for speculative prices. *Econometrica, 41*, 135–155.
46. Clayton, D. G. (1978). A model for association in bivariate life tables and its application in epidemiological studies of familial tendency in chronic disease incidence. *Biometrika, 65*, 141–151.
47. Coles, S., Currie, J., & Tawn, J. (1999). *Dependence measures for extreme value analysis*. Department of Mathematics and Statistics, Lancaster University working paper.
48. Cont, R. (2001). Empirical properties of asset returns: Stylized facts and statistical issue. *Quantitative Finance, 1*, 223–236.
49. Cont, R., & Tankov, P. (2004). *Financial Modelling with Jump Processes*. Boca Raton: Chapman & Hall.
50. Cook, R. D., & Johnson, M. E. (1981). A family of distributions for modeling non-elliptically symmetric multivariate data. *Journal of the Royal Statistical Society Series B, 43*, 210–218.
51. Cook, R. D., & Johnson, M. E. (1986). Generalized Burr-Pareto-logistic distributions with applications to a uranium exploration data set. *Technometrics, 28*, 123–131.
52. Cox, D. R., & Oakes, D. (1984). *Analysis of Survival Data*. London: Chapman and Hall.
53. Cox, J. C., Ingersoll, J., & Ross, S. (1985). A theory of the term structure of interest rates. *Econometrica, 53*, 385–407.
54. CreditMetrics Manual RiskMetrics Group available in www.riskmetrics.com (2007).
55. Dambis, K. E. (1965). On the decomposition of continuous submartingales. *Theory of Probability and Its Applications, 10*, 401–410.
56. Darsow, W. F., Nguyen, B., & Olsen, E. T. (1992). Copulas and Markov processes. *Illinois Journal of Mathematics, 36*, 600–642.
57. Davidson, R., & MacKinnon, J. (1993). *Estimation and Inference in Econometrics*. New York: Oxford University Press.
58. Davydov, Y. (1973). Mixing conditions for Markov chains. *Theory of probability and its Applications, 18*, 312–328.
59. de Haan, L. (1970). On regular variation and its application to the weak convergence of sample extremes. Mathematics Centrum Amsterdam.
60. de Haan, L., & Ferreira, A. (2006). *Extreme Value Theory: An Introduction*. New York: Springer.
61. de Haan, L., & Resnick, S. I. (1977). Limit theory for multivariate sample extremes. *Z. Wahrscheinlichkeitstheorie und Verw. Gebiete, 40*(4), 317–337.
62. Deheuvels, P. (1978). Caractérisation complète des Lois Extrèmes Multivariées et de la Convergence des Types Extrèmes. *Publications de l'Institut de Statistique de l'Université de Paris, 23*, 1–36.
63. Deheuvels, P. (1979). La function de dépendance empirique et ses propriétés. Un test non paramétriquen d'indépendace. *Academic Royal Belg Bull Cl Sci, 65*(5), 274–292.
64. Deheuvels, P. (1981). A non parametric test for independence. Université de Paris, Institut de Statistique.
65. Devroye, L. (1986). *Non-Uniform Random Variate Generation*. New York: Springer.
66. Diebold, F. X., Gunther, T., & Tay, A. S. (1998). Evaluating density forecasts with applications to financial risk management. *International Economic Review, 39*, 863–883.
67. Diebold, F. X., Hahn, J., & Tay, A. S. (1999). Multivariate density forecast evaluation and calibration in financial risk management. *Review of Economics and Statistics, 81*(4), 661–673.

68. Dobric, J., & Schmidt, F. (2002). Nonparametric estimation of the lower tail dependence λ_L in bivariate copula. *Journal of Applied Statistics, 32*(4), 387–407.
69. Drouet-Mari, D., & Kotz, S. (2001). *Correlation and Dependence*. London: Imperial College Press.
70. Dubins, L. E., & Schwarz, G. (1965). On continuous martingales. *Proceedings of the National Academy Sciences USA, 53*, 913–916.
71. Durante, F. (2009). Construction of non-exchangeable bivariate distribution functions. *Statistical Papers, 50*(2), 383–391.
72. Durante, F., Hofert, M., & Scherer, M. (2009). Multivariate hierarchical copulas with shocks. *Methodology and Computing in Applied Probability, 12*(4), 681–694.
73. Durante, F., Foschi, R., & Sarkoci, P. (2010). Distorted copulas: Constructions and tail dependence. *Communications in Statistics-Theory and Methods, 39*(12), 2288–2301.
74. Durbin, J., & Stuart, A. (1951). Inversions and rank correlations. *Journal of the Royal Statistical Society Series B, 2*, 303–309.
75. Durrleman, V., Nikeghbali, A., & Roncalli, T. (2000a). Copulas approximation and new families. Groupe de Recherche Opèrationelle, Crédit Lyonnais, working paper.
76. Durrleman, V., Nikeghbali, A., & Roncalli, T. (2000b). Which copula is the right one? Groupe de Recherche Opèrationelle, Crédit Lyonnais, working paper.
77. Efron, B., & Tibshirani, R. J. (1993). *An Introduction to Bootstrap*. London: Chapman & Hall.
78. El Karoui, N., Jeanblanc, M., & Jiao, Y. (2009). Successive default times. *working paper*.
79. Embrechts, P., & Hofert, M. (2011). A note on generalized inverses. submitted, available on http://www.math.ethz.ch/~baltes/ftp/inversesPEMH.pdf.
80. Embrechts, P., & Puccetti, G. (2006a). Bounds for functions of dependent risks. *Finance and Stochastics, 10*, 341–352.
81. Embrechts, P., & Puccetti, G. (2006b). Bounds for functions of multivariate risks. *Journal of Multivariate Analysis, 27*, 526–547.
82. Embrechts, P., & Puccetti, G. (2006c). Aggregating risk capital, with an application to operational risk. *Geneva Risk and Insurance Review, 31*, 71–90.
83. Embrechts, P., Kluppenberg, P., & Mikosch, T. (1997). *Modeling Extremal Event for Insurance and Finance*. Berlin: Springer.
84. Embrechts, P., McNeil, A., & Straumann, D. (2002). Correlation and dependence in risk management: Properties and pitfalls. In M. A. H. Dempster (Ed.), *Risk Management: Value at Risk and Beyond*. Cambridge: Cambridge University Press, pp. 176–223.
85. Embrechts, P., Hoing, A., & Juri, A. (2003). Using Copulae to bound the value-at-risk for functions of dependent risks. *Finance and Stochastics, 7*, 145–167.
86. Emmer, S., & Tasche, D. (2003). Calculating credit risk capital charges with the one-factor model. *working paper*.
87. Engle, R. F. (Ed.). (1996). *ARCH Selected Readings*. Oxford: Oxford University Press.
88. Engle, R. F. (2002). Dynamic conditional correlation. *Journal of Business & Economic Statistics, 20*(3), 339–350.
89. Engle, R. F. (2008). *Anticipating Correlations: A New Paradigm for Risk Management*. Princeton: Princeton University Press.
90. Engle, R. F., & Manganelli, S. (1999). CAViaR: Conditional Autoregressive Value at Risk by regression quantiles. UCSD Department of Economics, working paper.
91. Engle, R. F., & Russell, J. (1998). The autoregressive conditional duration model. *Econometrica, 66*, 1127–1162.
92. Engle, R. F., Ng, V., & Rothschild, M. (1990). Asset pricing with a Factor ARCH covariance structure: Empirical estimates for Treasury Bills. *Journal of Econometrics, 45*, 213–237.
93. Erb, C. B., Harvey, C. R., & Viskanta, T. E. (1994). Forecasting international equity correlations. *Financial Analysts Journal, 50*(6), 32–45.
94. Ethier, N. E., & Kurtz, T. G. (1985). *Markov Processes: Characterization and Convergence*. New York: Wiley.
95. Evans, M., Hastings, N., & Peacock, B. (1993). *Statistical Distributions*. New York: Wiley.

96. Feller, W. (1968). *An Introduction to Probability Theory and Its Applications*, Vol. I. New York: Wiley.
97. Feller, W. (1971). *An Introduction to Probability Theory and Its Applications*, Vol. II. New York: Wiley.
98. Ferguson, T. S. (1995). A class of symmetric bivariate uniform distributions. *Statistics Papers, 36*, 31–40.
99. Fermanian, J. D., & Wegkamp, M. (2004). Time dependent copulas. Preprint INSEE. France: Paris.
100. Finger, C. (Ed.). (2003). CreditGrades Technical Document.
101. Fisher, M., Kock, C., Schlouter, S., & Weigert, F. (2009). An empirical analysis of multivariate copula models. *Quantitative Finance, 9*(7), 839–854.
102. Florens, J. P., & Fougere, D. (1991). Non-causality in continuous time: applications to counting processes. Cahier du GREMAQ 91.b, Universite des Sciences Sociales de Toulouse.
103. Frank, M. J. (1979). On the simultaneous associativity of $F(x, y)$ and $x + y - F(x, y)$. *Aequationes Mathematicae, 19*, 194–226.
104. Frank, M. J., & Schweizer, B. (1979). On the duality of general infimal and supremal convolutions. *Rend Math, 12*, 1–23.
105. Frank, M. J., Nelsen, R., & Schweizer, B. (1987). Best-poss bounds for the distribution of a sum – a problem of Kolmogorov. *Probability Theory and Related Fields, 74*, 199–211.
106. Fréchet, M. (1935). Généralisations du théorème des probabilités totales. *Fundamenta Mathematicae, 25*, 379–387.
107. Fréchet, M. (1951). Sur le Tableaux de Corrélation dont les Marges sont données. *Annales de l'Université de Lyon, 9*, Sect. A, 53–77.
108. Fréchet, M. (1958). Remarques au sujet de la note précédente. *Comptes Rendus de l'Académie des Sciences Paris, 246*, 2719–2720.
109. Frey, R., & Mc Neil, A. J. (2001). Modeling dependent defaults. Department of Mathematics ETH Zurich, working paper.
110. Gagliardini, P., & Gouriéroux, C. (2007). An efficient nonparametric estimator for models with nonlinear dependence. *Journal of Econometrics, 137*, 189–229.
111. Gaudenzi, R. (2016). Loss Function di un portafoglio multivariato: un approccio Bayesiano. Master Thesis, University of Bologna.
112. Geluk, J. L., & de Haan, L. (1987). Regular variation, extensions and Tauberian Theorems. *CWI Tract. Stichting Mathematisch Centrum* vol.40, centrum voor Wiskunde en Informatica, Amsterdam.
113. Genest, C. (1987). Frank's family of bivariate distributions. *Biometrika, 74*, 549–555.
114. Genest, C., & MacKay, J. (1986). The joy of copulas: Bivariate distributions with uniform marginals. *American Statistician, 40*, 280–283.
115. Genest, C., & Neslehova, J. (2007). A primer on copulas for count data. *ASTIN Bulletin, 37*, 475–515.
116. Genest, C., Ghoudi, K., & Rivest, L. (1995). A semiparametric estimation procedure of dependence parameters in multivariate families of distributions. *Biometrika, 82*(3), 543–552.
117. Genest, C., Ghoudi, K., & Rivest, L. P. (1995). A semiparametric estimation procedure of dependence parameters in multivariate families of distributions. *Biometrika, 82*(3), 543–552.
118. Georges, P., Lamy, A., Nicolas, E., Quibel, G., & Roncalli, T., (2001). Multivariate survival modelling: A unified approach with copulas. Groupe de Recherche Opèrationelle, Crédit Lyonnais, working paper.
119. Ghysel, E., Gouriéroux, C., & Jasiak, J. (2004). Stochastic volatility duration models. *Journal of Econometrics, 119*, 413–433.
120. Gibbons, J. D. (1992). *Nonparametric Statistical Inference*. New York: Marcel Dekker Inc.
121. Gourieroux, C., & Monfort, A. (2002). Equidependence in qualitative and duration models with application to credit risk. CREST, working paper.
122. Gourieroux, C., & Robert, C. Y. (2006). Stochastic unit root models. *Econometric Theory, 22*, 1052–1090.
123. Gourieroux, C., Laurent, J. P., & Scaillet, O. (2000). Sensitivity analysis of values at risk. *Journal of Empirical Finance, 7*, 225–245.

124. Granger, C. J. W. (2003). Time series concepts for conditional distributions. *Oxford Bulletin for Economics and Statistics, 65*, 689–701.
125. Granger, C. J. W., Maasoumi, E., & Racine, J. (2004). A dependence metric for possibly nonlinear processes. *Journal of Time Series Analysis, 25*, 649–669.
126. Greco, S., Matarazzo, B., & Slowinski, R. (2001). Rought sets theory for multicriteria decision analysis. *European Journal Operational Researches, 129*(1), 1–47.
127. Gumbel, E. J. (1960). Bivariate exponential distributions. *Journal of the American Statistical Association, 55*, 698–707.
128. Härdle, W. (1990). *Applied Nonparametric Regression*. ESM Cambridge: Cambridge University Press.
129. Henriksson, R. D., & Merton, R. C. (1981). On Market Timing and investment performance. II. Statistical procedures for evaluating forecasting skills. *Journal of Business, 54*(4), 513–533.
130. Heston, S. L. (1993). A closed form solution for options with stochastic volatility with applications to bond and currency options. *Review of Financial Studies, 6*, 327–343.
131. Hiemstra, C., & Jones, J. D. (1994). Testing for linear and non-linear Grange causality in the stock price-volume relation. *Journal of Finance, 49*, 1639–1664.
132. Hoeffding, W. (1940). Masstabinvariante Korrelationstheorie. *Schriften des Mathematischen Instituts und des Instituts fur Angewandte Mathematik der Universitat Berlin, 5*, 179–233.
133. Hofert, M. (2010). Sampling nested Archimedean copulas with applications to CDO pricing. *Südwestdeutscher Verlag für Hochschulschriften AG & Co. KG*, Ph.D thesis.
134. Hofert, M. (2011a). Efficiently sampling nested Archimedean copulas. *Computational Statistics and Data Analysis, 55*, 57–70.
135. Hofert, M. (2011b). A stochastic representation and sampling algorithm for nested Archimedean copulas. *Journal of Statistical Computation and Simulation, 55*, 57–70.
136. Hofert, M., & Scherer, M. (2010). CDO pricing with nested Archimedean copulas. *Quantitative Finance, 10*. 775–787.
137. Hougaard, P. (1986). A class of multivariate failure time distributions. *Biometrika, 73*, 671–678.
138. Huang, Y., & Kou, G. (2014). A kernel entropy manifold learning approach for financial data analysis. *Decision Support Systems, 64*, 31–42.
139. Huang, W., & Prokhorov, A. (2008). Goodness-of-fit test for copulas. Revised and Resubmitted, *Econometric Reviews*.
140. Hull, J., & White, A. (1998). Value at Risk when daily changes in market variables are not normally distributed. *Journal of Derivatives, 5*(3), 9–19.
141. Hutchinson, T. P., & Lai, C. D. (1990). *Continuous Bivariate Distributions, Emphasising Applications*. Adelaide: Rumsby Scientific Publishing.
142. Ibragimov, R. (2005). Copula-based dependence characterization and modeling for time series. Harvard Institute of Economic Research, Discussion paper n. 2094.
143. Ibragimov, R. (2009). Copula based characterizations for higher-order Markov processes. *Econometric Theory, 25*(3), 819–846.
144. Ibragimov, R., & Lentzas, G. (2009). Copulas and long memory. Harvard Institute of Economic Research Discussion Paper No. 2160. Revised November 2009.
145. Jacod, J., & Shiriaev, A. N. (2003). *Limit Theorems For Stochastic Processes*, 2nd edn. Berlin: Springer.
146. Jaworski, P., Durante, F., Haerdle, W., & Rychlik T. (Eds.). (2010). *Copula Methods and Its Applications*. Volume 198 of Lecture Notes in Statistics-Proceedings. Berlin/Heidelberg: Springer.
147. Joag-Dev, K. (1984). Measures of dependence. In P. R. Krishnaiak (Ed.), *Handbook of Statistics*. New York/North-Holland: Elsevier, Vol. 4, pp. 79–88.
148. Joe, H. (1989). Relative entropy measures of multivariate dependence. *Journal of the American Statistical Association, 84*(405), 157–164.
149. Joe, H. (1990). Multivariate concordance. *Journal of Multivariate Analysis, 35*, 12–30.

150. Joe, H., & Hu, T. (1996a). Multivariate distributions from mixtures of max-infinitely divisible distributions. *Journal of Multivariate Analysis, 57*, 240–265.
151. Joe, H., & Xu, J. J. (1996b). The estimation method of inference functions for margins for multivariate models. Department of Statistics University of British Columbia, Technical report 166.
152. Jouanin, J. F., Rapuch, G., Riboulet, G., & Roncalli, T. (2001). Modelling dependence for credit derivatives with copulas. Groupe de Recherche Opèrationelle, Crédit Lyonnais working paper.
153. Kallsen, J., & Tankov, P. (2006). Characterization of dependence of multidimensional Lévy processes using Lévy copulas. *Journal of Multivariate Analysis, 97*, 1551–1572.
154. Kendall, M. G. (1938). A new measure of rank correlation. *Biometrika, 30*, 81–93.
155. Kendall, M. G. (1970). *Rank Correlation Methods*. London: Griffin.
156. Khoudraji, A. (1995). Contribution l'étude des copules et la modèlisation de valeurs extrêmes multivarièes. PhD Thesis, Universit Laval, Qubec.
157. Kimberling, C. H. (1974). A probabilistic interpretation of complete monotonicity. *Aequationes Mathematicae, 10*, 152–164.
158. Kimeldori, G., & Sampson, A. R. (1975). Uniform representation of bivariate distributions. *Communications in Statistics, 4*, 617–627.
159. Kim, G., Silvapulle, M., & Silvapulle, P. (2007). Comparison of semiparametric and parametric methods for estimating copulas. *Computational Statistics and Data Analysis, 51*, 2836–2850.
160. Tae-Hwan, K., & White, H. (2002). Estimation inference and testing for possibly misspecified quantile regression. In T. B. Fomby & R. C. Hill (Eds.). *Maximum Likelihood Estimation of Misspecified Models: Twenty Years Later* (Advances in Econometrics, Volume 17), Emerald Group Publishing Limited, pp. 107–132.
161. Klein, R. W., & Dubes, R. C. (1989). Experiments in projection and clustering by simulated annealing. *Pattern Recognition, 22*(2), 213–220.
162. Klement, E. P., Mesiar, R., & Pap, E. (2002). Invariant Copulas. *Kybernetica (Prague)*, 275–285.
163. Kluppelberg, C., & Resnick, S. (2008). The Pareto copula, aggregation of risks and Emperor's socks. *Journal of Applied Probabilities, 45*(1), 67–84.
164. Koenker, R., & Basset, Jr. G. (1978). Regression quantiles. *Econometrica, 46*(1), 33–50.
165. Koenker, R., & Park, J. B. (1996). An interior point algorithm for nonlinear quantile regression. *Journal of Econometrics, 71*(1), 265–283.
166. Koenker, R., & Xiao, Z. (2006). Quantile autoregression. *Journal of the American Statistical Association, 101*, 980–990.
167. Kolmogorov, A. N., & Rozanov, Y. A. (1960). On the strong mixing conditions for stationary Gaussian sequences. *Theory of probability and Its Applications, 5*, 204–207.
168. Konijn, H. S. (1959). "Positive and negative dependence of two random variables". *Sankhyà, 21*, 269–80.
169. Kou, G., Peng, Y., & Lu, C. (2014a). An MCDM approach to evaluate bank loan default models. *Technological and Economic Development of Economy, 20*(2), 278–297.
170. Kou, G., Peng, Y., & Wang, G. (2014b) Evaluation of clustering algorithms for financial risk analysis using MCDM methods. *Information Sciences, 275*, 1–12.
171. Lando, D. (2004). *Credit Risk Modeling: Theory and Applications*. Princeton: Princeton University Press.
172. Laurent, J. P., & Gregory, J. (2002). Basket Default Swaps, CDOs and Factor Copulas, October, BNP Paribas and University of Lyon working paper.
173. Lee, A. J. (1993). Generating random binary deviates having fixed marginal distributions and specified degrees of association. *American Statistician, 47*, 209–215.
174. Lehmann, E. (1966). Some concepts of dependence *Annals of Mathematical Statistics, 37*, 1137–1153.
175. Lehmann, E., & Casella, G. (1998). *Theory of Point Estimation*. New York: Springer.

176. Li, D. X. (2000). On default correlation: A copula function approach. *Journal of Fixed Income, 9*(4), 43–54.
177. Liebscher, E. (2008). Construction of asymmetric multivariate copulas. *Journal of Multivariate Analysis, 99*(10), 2234–2250.
178. Li, X., Mikusinski, P., Sherwood, H., & Taylor, M. D. (1997). On approximation of copulas. In V. Benes, J. Stepan, G. Dall'Aglio, S. Kotz, G. Salinetti (Eds.), *Distributions with Given Marginals and Moment Problems*. Dordrecht: Kluwer Academic Publishers.
179. Li, J., Feng, J., Sun, X., & Li, M. (2012). Risk integration mechanisms and approaches in banking industry. *International Journal of Information Technology & Decision Making, 11*(6), 1183–1213.
180. Lim, K. G., Chang, S., & Tsui, K. C. (2002). Defaultable debt pricing in multi-factor models. *International Journal of Theoretical and Applied Finance, 5*(8), 823–844.
181. Lindskog, F. (2000). Linear correlation estimation. RiskLab ETH working paper.
182. Lindskog, F., Mc Neil, A. J., & Schmock, U. (2001). A note on Kendall's tau for elliptical distribution. ETH Zurich, *working paper*.
183. Ling, C. H. (1965). Representation of associative functions. *Publicationes Mathematicae Debrecen, 12*, 189–212.
184. Livan, G., Novaes, M., & Vivo, P. (2018). *Introduction to Random Matrices: Theory and Practice*. Springer Briefs in Mathematical Physics.
185. Longin, F., & Solnik, B. (2001). Extreme correlation of international equity markets. *The Journal of Finance, 56*(2), 649–676.
186. Luciano, E., & Marena, M. (2002). Portfolio value at risk bounds. *International Transactions in Operational Research, 9*(5), 629–41.
187. Luciano, E., & Marena, M. (2003a). Value at risk bounds for portfolios of non-normal returns. In C. Zopoudinis (Ed.), *New Trends in Banking Management*. Berlin: Physica-Verlag, pp. 207–22.
188. Luciano, E., & Marena, M. (2003b). Copulae as a new tool in financial modelling. *Operational Research: An International Journal, 2*(2), 139–155.
189. Luciano, E., & Schoutens, W. (2006). A multivariate jump-driven financial asset model. *Quantitative Finance, 6*(5), 385–402.
190. Madan, D., & Seneta, E. (1998). The Variance Gamma (VG) model for share market returns. *Journal of Business, 63*(4), 511–524.
191. Mai, J. F. (2010). Extendibility of Marshall-Olkin distributions via Lévy subordinators and an application to portfolio credit risk. Ph.D thesis, Technische Universitat Munchen Zentrum Mathematik HVB-Stiftungsinstitut fur Finanzmathematik.
192. Mai, J. F., & Scherer, M. (2009a). A tractable multivariate default model based on a stochastic time-change. *International Journal of Theoretical and Applied Finance, 12*(2), 227–249.
193. Mai, J. F., & Scherer, M. (2009b). Pricing k-th to default swaps in a Lèvy-time framework. *Journal of Credit Risk, 5*(3), 55–70.
194. Mai, J. F., & Scherer, M. (2011). Reparameterizing Marshall-Olkin copulas with applications to sampling. *Journal of Statistical Computation & Simulation, 81*(1), 59–78.
195. Mai, J. F., & Scherer, M. (2017). *Simulating Copulas: Stochastic Models, Sampling Algorithms, and Applications*. Singapore: World Scientific.
196. Mai, J. F., Scherer, M., & Zagst, R. (2011). CIID default models and implied copulas. working paper.
197. Makarov, G. D. (1981). Estimates for the distribution function of a sum of two random variables when the marginal distributions are fixed. *Theory of Probability and Its Applications, 26*, 803–806.
198. Marchenko, V. A., & Pastur, L. A. (1967). Distribution of eigenvalues for some sets of random matrices. *Mathematics of the USSR-Sbornik, 1*(4), 507–536.
199. Martin, R., & Wilde, T. (2003). Unsystematic credit risk. *Risk, 15*(3), 26–32.
200. Mashal, R., & Zeevi, A., (2002) Beyond correlation: Extreme co-movements between financial assets. Columbia Business School, working paper.

201. McCullagh, P., & Nelder J. A. (1989). *Generalized Linear Models*. London: Chapman and Hall.
202. McNeil, A. J. (2008). Sampling nested Archimedean Copulas. *Journal of Statistical Computational Simulation, 78*(6), 567–581.
203. McNeil, A. J., Frey, R., & Embrechts, P. (2005). *Quantitative Risk Management. Concepts, Techniques and Tools*, Princeton Series in Finance. Princeton: Princeton University Press.
204. Mehta, M. L. (1991). *Random Matrices*, 3rd edn. New York: Academic.
205. Meneguzzo, D., & Vecchiato, W. (2000). "Improvement on Value at Risk measures by combining Conditional Autoregressive and Extreme Value approaches" *Review in International Business and Finance* Special Issue on Financial Risk and Financial Risk Management, Vol. 16, Cap. 12, 275–324, Elsevier Science Ltd. Apr 2002.
206. Meneguzzo, D., & Vecchiato, W. (2003). Copula sensitivity in Collateralized Debt Obligations and Basket Default Swaps. *The Journal of Future Markets, 24*(1), 37–70.
207. Merton, R. C. (1974). On the pricing of corporate debt: The risk structure of interest rates. *Journal of Finance, 29*, 449–470.
208. Merton, R. C. (1981). On market timing and investment performance. II. An equilibrium theory of value for market forecests. *Journal of Business, 54*(3), 363–406.
209. Mikusinski, P., Sherwood, H., & Taylor, M. D. (1992). Shuffles of min. *Stochastica, 13*, 61–74.
210. Monroe, I. (1978). Processes that can be embedded in Brownian motion. *Annals of Applied Probability, 6*(1), 42–56.
211. Moore, D. S., & Spruill, M. C. (1975). Unified large-sample theory of general chisquared statistics for tests of fit. *Annals of Statistics, 3*, 599–616.
212. Morgan, J. P. (2000). Guide to credit derivatives. *Risk*.
213. Mortensen, A. (2006). Semi-analytical valuation of basket credit derivatives in intensity-based models. *Journal of Derivatives, 13*, 8–26.
214. Mortensen, A. (2006). Semi-analytical valuation of basket credit derivatives in intensity-based models. *Journal of Derivatives, 13*, 8–26.
215. Moynihan, R., Schweizer, B., & Sklar, A. (1978). Inequalities: Among binary operations on probability distribution functions. *General Inequalities, 1*, 133–149.
216. Muller, A., & Scarsini, M. (2000). Some remarks on the supermodular order. *Journal of Multivariate Analysis, 73*(1), 107–119.
217. Nagpal, K., & Bahar, R. (2001). Measuring default correlation. *Risk, 14*, 129–132.
218. Nelsen, R. B. (1996). Nonparametric measures of multivariate association. In *Distibution with Fixed Marginals and Related Topics*. IMS Lecture Notes-Monograph Series, Vol. 28, pp. 223–232.
219. Oakes, D. (1982). A model for association in bivariate survival data. *The Journal of the Royal Statistical Society, Series B, 44*, 414–422.
220. Oakes, D. (1986). Semiparametric inference in a model for association in bivariate survival data. *Biometrika, 73*, 353–361.
221. O'Kane, D., & Livesey, M. (2004). Base correlation explained. *Quantitative Credit Research, Lehman Brothers, Q3*(4).
222. Papageorgiou, E., & Sircar, R. (2008). Multiscale intensity models and name grouping for valuation of multi-name credit derivatives. *Applied Mathematical Finance, 15*, 73–105.
223. Patton, A. J. (2001). *Modelling time-varying exchange rate dependence using the conditional copula*. UCSD Department of Economics, working paper.
224. Patton, A. J. (2006a). Modelling asymmetric exchange rate dependence. *International Economic Review, 47*(2), 527–556.
225. Patton, A. J. (2006b). Estimation of multivariate models for time series of possibly different lengths. *Journal of Applied Econometrics, 21*, 147–173.
226. Pearson, N. D., & Sun, T. S. (1994). Exploiting the conditional density in estimating the term structure: An application to the cox-ingersoll-ross model. *Journal of Finance, 49*, 1279–1304.
227. Plackett, R. L. (1965). A class of bivariate distributions. *Journal of the American Statistical Association, 60*, 516–522.

228. Prokhorov, A., & Schmidt, P. (2009). Likelihood-based estimation in a panel setting: Robustness, redundancy and validity of copulas. *Journal of Econometrics, 153*, 93–104.
229. Rau-Breedow, H. (2002). Credit Portfolio Modelling, Marginal Risk Contributions and Granularity Adjustment. working paper.
230. Rémillard, B. (2010). Goodness-of-fit tests for copulas of multivariate time series. SSRN Working Paper Series No. 1729982.
231. Rémillard, B., Papageorgiou, N., & Soustra, F. (2010). Dynamic copulas. Technical Report G-2010-18, Gerad. 1.
232. Rényi, A. (1959). On measures of dependence. *Acta Mathematica Academiae Scientiarum Hungaricae, 10*, 441–451.
233. Resnick, S. I. (1987). *Extreme values, Regular Variation and Point Processes*. New York: Springer.
234. Resnick, S. I. (2007). *Heavy Tail Phenomena: Probabilistic and Statistical Modeling*. Springer Series in Operations Research and Financial Engineering. New York: Springer.
235. Robinson, P. (1983). Nonparametric estimators for time series. *Journal of Time Series Analysis, 4*, 185–207.
236. Rockafellar, R. T., & Uryasev, S. (2000). Optimization of conditional value-at-risk. *Journal of Risk, 2*, 21–41.
237. Rockinger, M., & Jondeau, E. (2001). Conditional dependency of financial series: An application of copulas. Les Cahiers de Recherche 723, HEC Paris.
238. Roncalli, T. (2001). Copulas: A tool for dependence in finance. Groupe de Recherche Opèrationelle, Crédit Lyonnais, working paper.
239. Roncalli, T. (2002). Gestiondes Risques Multiples. Cours ENSAI de 3' année, Groupe de Recherche Opèrationelle, Crédit Lyonnais, working paper.
240. Roncalli, T. (2003). Gestion des Risques Multiples: Le Risque de Credit 2ème partie. Groupe de Recherche Opèrationelle, Crèdit Lyonnais, working paper.
241. Rosemberg, J. V. (2000). Nonparametric pricing of multivariate contingent claims. Stern School of Business, working paper.
242. Rousseuw, P., & Molenberghs, G. (1993). Transformations of non positive semidefinite correlation matrices. *Communications in Statisitcs Theory and Methods, 22*(4), 965–984.
243. Rouvinez, C. (2003). Diversity scoring for market value CDOs. *Risk*, 514–516.
244. Ruschendorf, L., Schweizer, B., & Taylor, M. D. (1996). *Distributions with Fixed Marginals and Related Topics*. Hayward: Institute of Mathematical Statistics.
245. Ruymgaart, F. H., & VAN ZUIJLEN, M. C. A. (1978). Asymptotic normality of multivariate-linear rank statistics in the non-i.i.d. case, *Annals of Statistics, 6*(3), 588–602.
246. Samorodnitsky, G., & Taqqu, M. S. (1995). *Stable Non-Gaussian Random Processes. Stochastic Models with Infinite Variance*. New York/London: Chapman and Hall.
247. Sato, K. (1999). *Lévy Processes and Infinitely Divisible Distributions*. Cambridge: Cambridge University Press.
248. Savu, C., & Trede, M. (2008). Goodness-of-fit tests for parametric families of Archimedean copulas. *Quantitative Finance, 8*, 109–116.
249. Scaillet, O. (2000). Nonparametric estimation of copulas for time series. IRES working paper.
250. Scarsini, M. (1984). On measures of concordance. *Stochastica, 8*, 201–218.
251. Schmid, F., & Schmidt, R. (2007). Multivariate conditional version of Spearman's rho and related measures of tail dependence. *Journal of Multivariate Analysis, 98*, Elsevier, 1123–1140.
252. Schmid, F., & Schmidt, R. (2006b) Bootstrapping Spearman's multivariate rho. In A. Rizzi & M. Vichi (Eds.), Proceedings of COMPSTAT 2006, pp. 759–766.
253. Schmid, F., & Schmidt, R. (2007). Multivariate extensions of Spearman's rho and related statistics. *Statistics and Probability Letters, 77*, 407–416.
254. Schmitz, V. (2003). Copulas and Stochastic Processes. Ph.D. dissertation, Aachen University.
255. Schönbucher, P., & Schubert, D. (2001). Copula-dependent default risk in intensity models. University of Bonn Germany, working paper.

256. Schweizer, B. (1991). Thirty years of copulas. In G. Dall'Aglio, S. Kotz, & G. Salinetti (Eds.), *Advances in Probability Distributions with Given Marginals*. Dordrecht: Kluwer Academic Publishers, pp. 13–50.
257. Schweizer, B., & Sklar, A. (1961). Associative functions and statistical triangle inequalities. *Publicationes Mathematicae Debrecen, 8*, 169–186.
258. Schweizer, B., & Sklar, A., (1974). Operations on distribution functions not derivable from operations on random variables. *Studia Mathematica, 52*, 43–52.
259. Schweizer, B., & Sklar, A. (1983). *Probabilistic Metric Spaces*. New York: Elsevier Science Publishing Co.
260. Schweizer, B., & Wolff, E. (1976). Sur une Mesure de Dépendance pour les Variables Aléatoires. *Comptes Rendus de l'Académie des Sciences de Paris, 283*, 659–661.
261. Schweizer, B., & Wolff, E. (1981). On non parametric measures of dependence for random variables. *Annals of Statistics, 9*, 879–885.
262. Scott, D. (1992). *Multivariate Density Estimation: Theory and Practice and Visualization*. New York: Wiley.
263. Serfling, R. J. (1980). *Approximation Theorems of Mathematical Statistics*. New York: Wiley.
264. Shao, J. (1999). *Mathematical Statistics*. New York: Springer.
265. Shao, J., & Tu, D. (1995). *The Jacknife and Bootstrap*. New York: Springer.
266. Silvapulle, P., & Moosa, I. (1999). The relationship between spot and futures prices: Evidence from the crude oil market. *Journal of Futures Markets, 19*, 147–167.
267. Silverman, B. (1986). *Density Estimation for Statistics and Data Analysis*. London: Chapman-Hall.
268. Sklar, A. (1973). Random variables, distribution functions, and copulas. *Kybernetica, 9*, 449–460.
269. Sklar, A. (1996). Random variables, distribution functions, and copulas a personal look backward and forward. In L. Ruschendorf, B. Schweizer, & M. D. Taylor (Eds.), *Distributions with Fixed Marginals and Related Topics*. Hayward: Institute of Mathematical Statistics, pp. 1–14.
270. Sklar, A. (1998). personal communication.
271. Song, P. (2000). Multivariate dispersion models generated from Gaussian copula. forthcoming in *Scandinavian Journal of Statistics*.
272. Stulz, R. M. (1982). Options on the minimum and maximum of two risky assets. *Journal of Finance, 10*, 161–185.
273. Tankov, P. (2003). Dependence structure of spectrally positive multidimensional Lévy processes. working paper.
274. Tasche, D. (1999). Risk Contributions and Performance Measurement. working paper.
275. Thorin, O. (1977). On the infinite divisibility of the lognormal distribution. *Scandinavian Actuarial Journal, 3*, 121–148. http://dx.doi.org/10.1080/03461238.1977.10405635.
276. Tjostheim, D. (1996). Measures of dependence and tests of indipendence. *Statistics, 28*, 249–284.
277. Valdez, E. A., & Xiao, Y. (2011). On the distortion of a copula and its margins. *Scandinavian Actuarial Journal, 4*, 292–317.
278. Van der Goorberg R. Genest, C., & Werker, B. (2005). Bivariate option pricing using dynamic copula models. *Insurance, Mathematics and Economics, 37*, 101–114.
279. Vidozzi, A. (2009). Two essays in mathematical finance. Doctoral dissertation, Illinois Institute of Technology.
280. Vidozzi, L. (2009). Two essays on multivariate stochastic processes and applications to credit risk modeling. Doctoral dissertation, Illinois Institute of Technology.
281. Wang, S. S. (1998). Aggregation of correlated risk portfolios: models & algorithms. CAS Committee on Theory of Risk, working paper.
282. White, H. (1984). *Asymptotic Theory for Econometricians*. New York: Academic.
283. Wigner, E. (1955). Characteristic vectors of bordered matrices with infinite dimensions. *Annals of Mathematics, 62*(3), 548–564.
284. Wilde, T. (2002). Probing granularity. *Risk, 14*(8), 103–106.

285. Williamson, R. C. (1989). Probabilistic arithmetic. Ph.D Thesis, University of Queensland.
286. Williamson, R. C., & Downs, T. (1990). Probabilistic arithmetic I: Numerical methods for calculating convolutions and dependency bounds. *International Journal of Approximate Reasoning, 4*, 89–158.
287. Wolff, E. F. (1981). N-dimensional measures of dependence. *Stochastica, 4*, 175–188.
288. Yu, J., & Phillips, P. C. B. (2001). Gaussian approach for continuous time models of the short term interest rate. *Econometrics Journal, 0*, 1–17.

The manufacturer's authorised representative in the EU is Springer
Nature Customer Service Centre GmbH, Europaplatz 3, 69115 Heidelberg,
Germany. If you have any concerns regarding our products, please
contact ProductSafety@springernature.com

Printed and bound by CPI Group (UK) Ltd, Croydon, CR0 4YY
28/09/2026
02224091-0002